D0216527

BT COMMUNICATIONS TECHNOLOGY SERIES 8

Location and Personalisation: Delivering Online and Mobility Services

Other volumes in this series:

Volume 1 **Carrier-scale IP networks: designing and operating Internet networks** P. Willis (Editor)
Volume 2 **Future mobile networks: 3G and beyond** A. Clapton (Editor)
Volume 3 **Voice over IP: systems and solutions** R. Swale (Editor)
Volume 4 **Internet and wireless security** R. Temple and J. Regnault (Editors)
Volume 5 **Broadband applications and the digital home** J. Turnbull and S. Garrett (Editors)
Volume 6 **Telecommunications network modelling, planning and design** S. Evans (Editor)
Volume 7 **Telecommunications performance engineering** R. Ackerley (Editor)

Location and Personalisation: Delivering Online and Mobility Services

Edited by
Daniel Ralph
and
Stephen Searby

The Institution of Electrical Engineers

Published by: The Institution of Electrical Engineers, London, United Kingdom

The Institution of Electrical Engineers,
Michael Faraday House,
Six Hills Way, Stevenage,
Herts. SG1 2AY, United Kingdom

www.iee.org

British Library Cataloguing in Publication Data

A catalogue record for this product is available from the British Library

ISBN 0 86341 338 2

Typeset in the UK by Bowne Global Solutions Ltd, Ipswich, Suffolk
Printed in the UK by T J International, Padstow, Cornwall

CONTENTS

	Preface	ix
	Introduction	xi
	Contributors	xv
1	**Personalisation — an Overview of its Use and Potential**	1
	S Searby	
	1.1 Introduction	1
	1.2 Success Factors	1
	1.3 Categories of Personalisation	2
	1.4 Privacy	5
	1.5 The Personal Profile	6
	1.6 Scenario — James the Architect	8
	1.7 Summary	11
2	**An Overview of Location-Based Services**	15
	T D'Roza and G Bilchev	
	2.1 Introduction	15
	2.2 Positioning Technology	15
	2.3 Co-ordinate Systems	20
	2.4 Representing Location in Applications	20
	2.5 Data Formats and Standards	21
	2.6 Applications	24
	2.7 Summary	28
3	**Locating Calls to the Emergency Services**	31
	P H Salmon	
	3.1 Introduction	31
	3.2 The Importance of Location in Emergencies	31
	3.3 Public Safety Answering Point (PSAP)	32
	3.4 The US Federal Communications Commission (FCC) 911 Mandate	33
	3.5 European Emergency Location Initiatives	34
	3.6 Co-ordination Group on Access to Location Information by Emergency Services (CGALIES)	35

	3.7	European Emergency Service Requirements	35
	3.8	Mobile Location Technology for Emergency Services	36
	3.9	Standards	39
	3.10	Location Privacy	40
	3.11	Geographic Information Systems	41
	3.12	Summary	42
4	**Location-Based Services — an Overview of the Standards**		43
	P M Adams, G W B Ashwell and R Baxter		
	4.1	Introduction	43
	4.2	Evolution of the Framework for the Development of LBS Standards	43
	4.3	LBS Work in 3GPP on UMTS Standards	49
	4.4	Work in the WAP Forum on Location-Based Services	55
	4.5	Summary	58
5	**Profiling — Technology**		59
	R Newbould and R Collingridge		
	5.1	Introduction	59
	5.2	Drivers	60
	5.3	Standards and Technologies	64
	5.4	The Government User Identity for Europe (GUIDE)	66
	5.5	Example Applications	76
	5.6	Summary	78
6	**Service Personalisation and Beyond**		81
	G De Zen		
	6.1	Introduction	81
	6.2	Service Personalisation	81
	6.3	User Profiling	85
	6.4	Presence Service	88
	6.5	Service Personalisation Process	89
	6.6	Summary	90
7	**Profiles — Analysis and Behaviour**		93
	M Crossley, N J Kings and J R Scott		
	7.1	Introduction	93
	7.2	History	94
	7.3	The Use of Profiling	97
	7.4	Understanding Focus	105
	7.5	Delivering Relevance	109
	7.6	Summary	111

8	**Device Personalisation — Where Content Meets Device**	115
	S Hoh, S Gillies and M R Gardner	
	8.1 Introduction	115
	8.2 Client Capabilities Recognition	116
	8.3 The Client Profile	117
	8.4 The Standards — Binding the Profile	119
	8.5 Content Negotiation — the Experiment	121
	8.6 Summary	125
9	**Multilingual Information Technology**	129
	S Appleby	
	9.1 Introduction	129
	9.2 Localisation and Internationalisation	130
	9.3 Text Display	133
	9.4 Machine Translation	136
	9.5 Summary	143
10	**Smart Tools for Content Synthesis**	147
	M Russ and D Williams	
	10.1 Introduction	147
	10.2 Understanding the Media Industry	147
	10.3 Experiments in the Creation of Personalised Television	156
	10.4 Results	156
	10.5 Next Steps	159
	10.6 Summary	159
11	**Personalised Advertising — Exploiting the Distributed User Profile**	161
	G Bilchev and D Marston	
	11.1 Introduction	161
	11.2 On-Line Personalised Advertising — Overview	162
	11.3 Models of On-Line Advertising	162
	11.4 Distributed Profile Advertising	164
	11.5 Performance Metrics for Distributed Profile Advertising	168
	11.6 Privacy in Advertising	169
	11.7 Future Trends and Summary	171
12	**Personalisation and Web Communities**	173
	S Case, M Thint, T Ohtani and S Hare	
	12.1 Introduction	173
	12.2 System Components	174
	12.3 System Integration	178
	12.4 Summary	183

13 **Location Information from the Cellular Network — an Overview** 185
W Millar
13.1 Introduction 185
13.2 Overview of Location-Based Technology 186
13.3 The Mobile Location Computer 188
13.4 MLC Accuracy 188
13.5 Applications 192
13.6 Future Directions 194
13.7 Summary 195

14 **A Multi-Agent System to Support Location-Based Group Decision Making in Mobile Teams** 197
H L Lee, M A Buckland and J W Shepherdson
14.1 Introduction 197
14.2 Literature Review 198
14.3 mPower — an Asynchronous GDSS for Mobile Teams 199
14.4 Illustrative Example 203
14.5 Discussion 207
14.6 Summary 208

15 **A Million Segments of One — How Personal Should Customer Relationship Management Get?** 211
N J Millard
15.1 Introduction 211
15.2 How Personal Should CRM Get? 212
15.3 The Practice of Personalisation and What It Means to Customers 214
15.4 Future Directions for Personalised CRM 219
15.5 Summary 220

Index 223

PREFACE

I am delighted to have the opportunity to introduce this book on location and personalisation, the seventh in the joint IEE/BT Communications Technology Series. Right now, BT, like many businesses, is facing some important challenges, including enhancing the experience for its customers and extending its business into new areas that will boost revenues.

Personalisation and location technologies are vitally important to achieving these objectives, as you will see from the chapters that follow. These technologies allow customers to receive an experience that is closely matched to their personal needs and to interact with an organisation in a friendly and efficient way, using whatever channel or device is most convenient for them. What's more, they can use BT's services to make their businesses more efficient through use of advanced technologies, such as knowledge and community tools and a range of mobility applications.

Broadband is set to change the lives for people both at home and at work. The applications that it unleashes will have the user at their core and, as you read through this book, you will gain an insight into some of the innovative technologies that BT is able to bring to the market-place.

I hope you find the contents of this book both interesting and stimulating, as well as providing encouragement for all organisations and businesses to begin to share in this growing market.

Pierre Danon
Chief Executive BT Retail

INTRODUCTION

New network technologies are often challenged with the question: "What is the 'killer application'?" and, similarly, new application technologies face the question: "Where will this be used?" We won't pretend to know the answers to these questions but, in pulling together the chapters for this book, we have tried to take a pragmatic view of two particular application technologies — location and personalisation — at a time when the world is charging towards the new network technologies of broadband and 3G.

Personalisation, although founded in the world of targeted marketing and on-line sales, has proven to be a key success factor in mobile telephones but not always so successful in the Internet world. Location-based services, meanwhile, were touted as the 'killer application' for upcoming and next-generation mobile data networks, but this has not yet proved to be the case. The limited display and on-the-move use of mobile devices lends itself to requesting services based on your current or intended location and the display of information relevant to your surroundings and personal requirements. Until recently, access to location data has been restricted, but with the release of this information to application developers — providing that privacy is not abused and that billing models can be sustained — there is huge potential revenue for application developers and network operators.

The plethora of applications developed are likely to be supported by a revenue share model and, with 45 million mobile users in the UK alone, of which at least 2 million already pay for text information services, it could prove to be the catalyst which increases the usage of those supplementary services that utilise location. This has obvious applications for consumer services, such as a colour map of your location delivered using the multimedia message service. For the corporate market, the use of location information in bespoke services has been common for a number of years — including vehicle-tracking for asset management and security purposes.

Personalisation is a technology that has passed through phases of popularity and dismissal, but much of this has been based on its application to on-line shopping and targeted advertising or in customisation — really a very limited aspect of personalisation. Questions have been raised about whether the money invested in the technology will be recovered and privacy issues have raised fears that information collected will be abused. The impact of personalisation on choice gives cause for some debate. What if the personalised content is not refined enough or indeed is too aggressive in removing options in the pursuit of the single perfect choice for the user? Can the level of intelligence be achieved to second-guess the whim of the user wanting to try something new?

The reality, though, is that personalisation has a much wider scope than many realise, having particular application in customer relationships, service integration and knowledge management. While personalisation as a technology is not a 'killer application' it certainly provides the foundations to deliver information that is important to you, while on the move. The context awareness that can be gained from assessing personal data regarding the user's diary, preferences and location, combined with public-domain information, such as weather or travel updates, is truly powerful. It is this synergy that has driven the subject matter for this book.

Going into more detail, Chapter 1 introduces personalisation and seeks not only to dispel the misunderstandings about what it can do, but also to highlight the opportunities it can bring across a wide range of services. It explores the powerful combination of a mobile device and personalised information, but also highlights the need for privacy controls and the platform intelligence required to deliver real benefit to the user.

Complementing this introduction, Chapter 2 provides an overview of the location and positioning systems that are enabling location-based services to be delivered now and into the future. Expanding on this, it also examines the format of location data and the application scenarios that make use of it.

The scenario of safety and security is explored in greater detail in Chapter 3. The significant increase in the number of emergency calls from mobile users that are unable to state their location has led to regulation by the FCC in the USA. The chapter introduces the thought process and technology limitations supporting discussion of a similar mandate in the EU.

The impact and importance of standards in the mobile arena is outlined in Chapter 4. Here the ongoing standardisation efforts of location-based services through 3GPP is assessed in terms of positioning technologies and other standards efforts such as the Open Mobile Alliance and Location Interoperability Forum.

Returning to usage of personal information, including location in a user profile, Chapter 5 discusses the major industry technologies — W3C P3P, Microsoft .Net, and the Liberty Alliance. BT's profile hosting technology is used as a case study to demonstrate potential usage scenarios.

The topic of user profiles is extended in Chapter 6, where service personalisation is discussed in the context of mobility. A great deal of standardisation effort has been undertaken to develop concepts such as the virtual home environment, the user agent profile, and the generic user profile, all of which are covered in this chapter.

Following the collection of personal information through to one obvious conclusion, the use of knowledge management techniques to manipulate, extrapolate and reproduce the relevant choices for the user is fundamental to realising the value of the user profile. Chapter 7 examines the techniques that form the armoury of the knowledge management tool-kit.

The importance of terminal capability in assisting the delivery of personalised content should not be underestimated. The required characteristics range from screen size and bandwidth, to terminal type — PDA or Web tablet. Chapter 8 addresses this subject and explains the options available to determine key characteristics of the connecting client.

The display of content is extended within Chapter 9, which looks at European and Far Eastern languages and their machine-based translation. It concludes with an outline of BT's research into 'data-oriented' translation, which utilises examples to allow the system to learn.

The topic of content creation and its formatting is tackled in Chapter 10. This brings about the prospect of a TV programme being built around a template with a presenter you particularly like!

Over the following two chapters, the implementation of personalisation techniques will be discussed as applied to specific scenarios, since it is only through the development and testing of new services that a greater understanding of appropriate usage will be achieved.

Using distributed profile information for on-line advertising is the subject presented in the first of these two chapters, Chapter 11. The ability to deliver personalised content, particularly advertising, without compromising the user's privacy is vital in maintaining not just revenue but trust between supplier and customer.

On-line or Web communities have been another key driver of the Internet, largely based on people's leisure pursuits. Bringing these communities of interest together in a corporate context has been a greater hurdle and is the subject of the second of the two chapters. Chapter 12 describes the implementation of a workflow and collaboration middleware platform named Pl@za, which demonstrates this ability. It uses intelligent agent technology both to deliver personalised news information and to seek out appropriate corporate experts to help with specific problems that may be facing the group.

The next two chapters are case studies that focus on the delivery of services using location and personalisation.

Firstly Chapter 13 covers the integration of location information to deliver business advantage and tackles corporate applications as these have proved to be the most significant business cases for deployment.

The use of intelligent agents deployed in a location-based support system for a closed user group is the topic of the second case study, in Chapter 14. This identifies the workflow for a mobile group using PDA devices to schedule and communicate work priorities. Taking this proposition to a field trial is invaluable in proving assertions of usage and usefulness.

In conclusion, Chapter 15 provides the final word on the use of personalisation in the relationship with the most important element in the system — the customer. It is clear that trust is earned and vital in providing the channel for personal information to flow — the provision of which must be carefully guarded and implemented.

Several case studies give insight to the successful behaviours in personalising a customer experience.

This book feels like a journey that has yet to be completed — the potential of these technologies has yet to be realised, both in terms of usage and successful business models.

And finally, we hope this book gives you both an insight into the technology capabilities and limitations surrounding its title, and also sufficient detail for you to be able to visualise the next steps being taken to realise the future.

Daniel Ralph
Mobile Technologies, BT Exact
daniel.ralph@bt.com

Stephen Searby
Multimedia Content Solutions, BT Exact
stephen.searby@bt.com

CONTRIBUTORS

P M Adams, Mobile Standards Management, BT Exact, Adastral Park

S Appleby, Language Technology Research, BT Exact, Adastral Park

G W B Ashwell, Mobile Standards and Projects, BT Exact, Adastral Park

R Baxter, Mobile Applications, BT Exact, Adastral Park

G Bilchev, Next Generation Web Research, BT Exact, Adastral Park

M A Buckland, Future Technology Research, BT Exact, Adastral Park

S Case, Computational Intelligence Research, BT Exact, Adastral Park

R Collingridge, Personalisation and Profile Hosting, BT Exact, Adastral Park

M Crossley, Knowledge Management Research, BT Exact, Adastral Park

G De Zen, Mobile Service Personalisation, Siemens Mobile Radio, Milan

T D'Roza, formerly Location-Based Services, BT Exact, Adastral Park

M R Gardner, Multinetwork Customer Solutions, Chimera, University of Essex

S Gillies, Wireless Application Development, BT Exact, Adastral Park

S Hare, Support Engineer, Fujitsu Teamware, UK

S Hoh, Personalisation Research, BT Asian Research Centre, Malaysia

N J Kings, Knowledge Communities, BT Exact, Adastral Park

H L Lee, Intelligent Systems Research, BT Exact, Adastral Park

D Marston, eBusiness Development, BT Exact, Adastral Park

W Millar, Mobile and Intranet Applications, BT Global Services, Adastral Park

N J Millard, Customer Interaction Management, BT Exact, Adastral Park

R Newbould, Web Services and Personalisation, BT Exact, Adastral Park

T Ohtani, Web Services Middleware, Fujitsu Laboratories, Japan

M Russ, Technology Analyst, BT Exact, Adastral Park

P H Salmon, Mobile Systems Engineering, BT Exact, Adastral Park

J R Scott, Virtual Learning Environments, Chimera, University of Essex

S Searby, Multimedia Content Solutions, BT Exact, Adastral Park

J W Shepherdson, Intelligent Systems, BT Exact, Adastral Park

M Thint, Technical Specialist, BT Global Services, Reston

D Williams, Creative Technology Research, BT Exact, Adastral Park

1

PERSONALISATION — AN OVERVIEW OF ITS USE AND POTENTIAL

S Searby

1.1 Introduction

'Whenever something is modified in its configuration or behaviour by information about the user, this is personalisation.'

It is hard to imagine a much broader definition of personalisation than the one above. However, the choice is deliberate, as the subject has been clouded by many pre-conceptions and market 'hype'. This chapter explains how the use of personalisation can now be applied in four distinct ways and, by examples and comparisons, highlights some of the important benefits and business opportunities offered.

Central to personalisation is the concept of a user profile and this has received considerable attention both as a means of attracting and locking users into particular services, and as a controversial topic when concerns about personal privacy are raised. This has, consequently, been an important subject for research and some of the technology resulting from that is described here.

1.2 Success Factors

A useful everyday example that highlights why personalisation is important is the mobile telephone. To show the significance, it is necessary to examine some of the success factors involved.

It is clear that people use mobile telephones in many situations where they could use a fixed line. For example, when in an unfamiliar office environment, people will often use their mobile even if there are fixed-line telephones present on the desks. This is often because the mobile is holding a personal database of contact numbers allowing the user to make calls without looking up or keying numbers. There is also, however, a significant percentage of the population (6 per cent of households in the UK [1]) who only have a mobile and do not subscribe to fixed line telephony services. People who share accommodation, or who tend to use temporary accommodation for a large proportion of the time (e.g. students), may account for some of this.

Mobile telephones are very personal devices. They can hold an individual's personal contact information, they fit in the pocket and they are taken almost everywhere. The frequency with which people customise the appearance and the sound of the mobile provide further evidence that they are treated as a personal fashion statement. Functionally, as well, they provide a communications medium for contacting a person rather than a location (such as a desk or a house) or an organisation.

1.3 Categories of Personalisation

Personalisation applications can currently be grouped into four areas — targeted marketing, customer relationships, service integration, and knowledge management. This provides a means of identifying the value proposition and also makes it easier to separate out some of the technology market sectors.

1.3.1 Targeted Marketing/Advertising

This area is centred on generating increased revenue mainly in the consumer market. It is widely recognised that the Internet has shifted the power-base towards the consumer. The consumer can access many suppliers to get the best deal and, with such a wide choice available to potential customers, it is now becoming necessary to do more to retain their attention.

That is not to say that personalisation techniques could restrict choice (although this is possible if the only index and search capabilities accessible to a user were under the control of individual service providers — as might occur on a digital TV network), but they could allow the shopping experience to be less frustrating and also to tempt customers with goods that are most likely to interest them.

One opportunity to apply personalisation to on-line retailing is in the exploitation of minority or niche markets.

The distribution in Fig 1.1 follows Zipf's law [2] (frequency $\propto 1/\text{rank}^a$, where $a \approx 1$) and applies to many real-world examples, such as borrowing books from libraries, video rentals, etc. The important thing to note is that the area under the tail of the curve is very significant. If applied to sales revenues from the products in a market, it shows that serving the wide range of niche requirements can be more lucrative than providing the lead product to the majority. Profits to be made on niche markets are likely to be greater since there is less competition. Personalisation has the capability to create the impression of delivering a niche product to each user.

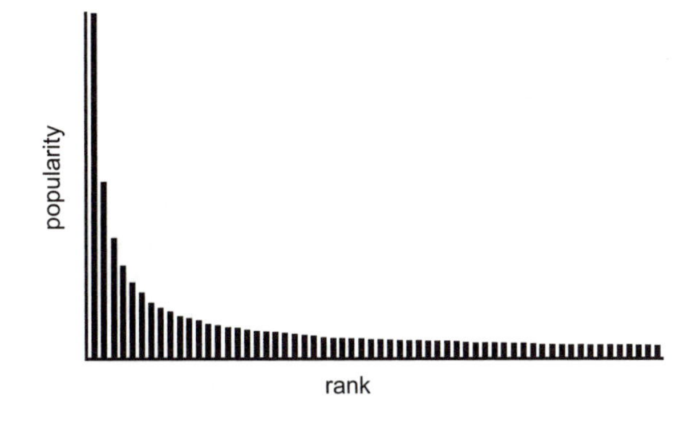

Fig 1.1 Zipf's popularity distribution profile.

Another opportunity from personalisation is the ability to exploit affinities [3]. A user who buys product *A* may be very likely to buy product *B* as well. This is based on tracking associated purchases such as beer with pizza.

Collaborative filtering [4] exploits similar behaviours by monitoring the patterns of groups of users and then trying to match a user with an appropriate group. Recommendations can be more sophisticated providing a detailed profile can be built up of users' purchasing patterns.

Loyalty cards enable personalised marketing techniques to be operated. The retailer can use the card to track what customers purchase and then send vouchers to match their interests or lifestyle to encourage the purchase of related products. If on-line applications can develop similar techniques to analyse users, then they could become very effective.

It is important to note, however, that there has been some scepticism about whether profiling is effective and how much additional revenue is gained from its use. Businesses have invested large amounts of money in the software to support it over recent years and many have found it difficult to measure the benefit [5, 6]. This may not be a fundamental problem but could simply be identifying the need for more sophisticated analysis techniques and a broader linkage of customer activity, i.e. the need to track behaviour across multiple services.

1.3.2 Customer Relationships

The term 'customer relationship management' (CRM) has often been used in this context. It represents more than just a technology approach but is a change of business philosophy. Rather than building an organisational structure and business processes around individual product or service groupings, a customer-centric approach is taken. Chapter 15 goes into more detail on this subject.

The intention is to provide improvements for the customer experience so that he or she is always recognised as the same individual regardless of the channel used to contact the organisation. In addition, system integration is applied to ensure that all communications channels and operational systems link to a common user account profile.

Consequently, as well as driving towards greater customer loyalty, the provision of multiple communications channels, including on-line access, can drive significant cost reductions in the provision of customer support. For the support cost reductions to be really effective the quality of the user experience through automated systems needs to be excellent. The ability to interface through a range of devices and in multiple languages needs to be considered as well.

Reductions in support costs can be measured in real businesses, but the benefits in terms of loyalty are less easy to quantify. There has been some doubt cast over whether consumers are looking for relationships with organisations:

> 'In a transaction-based, product-centric business model, buyer and seller are adversaries, no matter how much the seller may try not to act the part.' [7]

1.3.3 Service Integration

A symptom of the drive to launch new products and services quickly, combined with organisational structures centred around products and services, has been the implementation of 'stove-pipe' systems. These often have their own customer database and single access channel. Situations can arise where a voice portal may have no integration with an on-line service despite supplying the same content from the same organisation.

Overcoming this kind of legacy problem can prove extremely costly, as it is likely to lead to large system integration activities or the replacement with unified service platforms. The use of personalisation technology can offer a lighter weight solution in many cases.

A personal profile — discussed in detail later in this chapter — can provide a common reference point for certain data used to deliver applications. For example, multiple customer identities can be logically linked to a single profile identity. In addition, context information and personal preferences can be stored so that separate applications can share data. The profile can even facilitate authentication and single sign-on. The use of the profile for transfer of context across services can be illustrated in a simple example — a young man accesses a cinema information service via text messaging on his mobile telephone. The application checks his location using the cell information supplied by his mobile network provider and generates a list of recommendations. In addition to returning the text message, it can also store the details of the nearest cinema in his personal profile. If he requires directions to the cinema he can call a voice portal number (provided in the message) and that application can access his profile to check his most recent request. He can then be given spoken instructions to find the cinema without the need to re-specify his requirements to the voice portal.

To take advantage of the personal profile an agreed interface (API) must be made available to the applications. The recent growth in interest in Web Services [8], as an integration technique, provides an ideal opportunity for the personal profile to become a key component across many applications.

1.3.4 Knowledge Management

Trends in the business environment are placing enormous pressure on users to manage an ever-increasing workload. Business users' demands to access information, while away from the office, and to process it, will drive the emergence of services to filter and target information to match the user's interests and the particular task in hand.

Some of the key features in this area include:

- responding to user context;
- matching information to interests and preferences;
- sharing information with other appropriate users in a community.

User context can include the geographic location, the type of device and access mechanism in use and the user's current role (e.g. at work, at home).

Matching to both interests and context provides clear benefits when searching for information [9]. Indeed, the interests and context can be very dynamic and may need to be adapted many times during a typical day as the user switches between projects and activities.

Community tools are becoming increasingly important to allow people to collaborate more efficiently — especially when their working environment is spread over multiple locations. This is not just relevant in a business context, though, as there are opportunities for consumer services offering collaboration on-line [10] and many claim to be generating significant revenues from their membership.

1.4 Privacy

Successful personalisation is dependent upon the establishment of a trust relationship between users (providing detailed and accurate data) and businesses (employing supplied information in a responsible way). User experiences with both on-line and off-line services, such as spamming with inappropriate material, unsolicited contact, and security concerns over how information is stored, have undermined user confidence and increased their reluctance to impart the necessary data.

The potential costs can be illustrated by an example that was reported in the press in January 2000 [11] when an on-line advertisement agency paid $1.7bn for one of the USA's largest databases of off-line customer data. The deal would allow the companies to combine on-line and off-line customer information and improve targeted marketing.

There was an ensuing backlash from horrified customers that led to a Federal Trade Commission investigation. Consumer lawsuits were filed and major partners distanced themselves and it was necessary for the CEO to issue an apology. Furthermore, all of this contributed to a very significant drop in the company's share price.

Senior privacy experts predict that users will become more active and have greater control over the release of their information [12]. Consequently businesses will need to build a trust relationship with their customers and a clear, documented privacy policy is essential.

In some cases the mere fact that an eBusiness can welcome a user by name, accurately predict what a user might want, or present a user returning to an Internet site with their earlier abandoned shopping trolley, may alert the public to issues of privacy.

In order to allay user concerns over data misuse, the infomediary businesses, which store and broker information between users and service providers, will need to provide tight control of the personal data they hold. This will ultimately result in the privacy and protection of information being under the control of the end user.

Initiatives like the platform for privacy preferences (P3P) project [13] are emerging as industry standards offering users more control over how their personal information is employed by Internet sites.

1.5 The Personal Profile

In order to deliver personalised content it is necessary to link three basic sources of information. The first is the content itself, but this content must include metadata (information about the content such as ownership details, formatting information, copyright and contractual data, etc) that defines particular characteristics such as the topics contained in the content. The second is the user profile. This defines a range of information about the user such as interests, context, location, etc. The third key element is a set of business rules that defines how an application, like a Web server, must use the profile, in combination with the metadata, to determine what and how to present the content.

Figure 1.2 shows the logical architecture at a schematic level. The application takes content from the store and presents it in a format that is matched to the end user's access device. This is normally achieved through the use of templates or scripts which, having determined the device type, can generate an output stream that has the right wrapper, e.g. HTML for a PC client or WML for a mobile telephone.

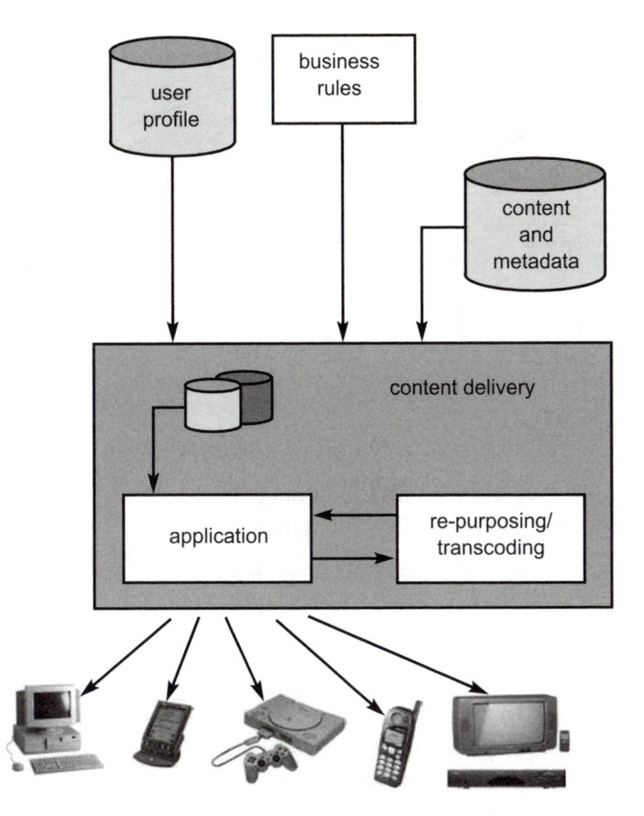

Fig 1.2 Key data sources required for personalisation.

Content is extracted from the store by the application and so can be dynamic, i.e. the information delivered will not be the same every time an access is made. Hence the application can deliver rapidly changing information, such as news, and can also provide customisation and personalisation to meet the individual user needs.

The personal user profile is a key element that must have a well-defined structure and API to allow multiple applications and services to make use of it. Figure 1.3 illustrates the typical elements within it but it is important to note that it must be extensible.

The profile can hold the standard personal data to allow forms to be pre-populated with address details, etc. It allows bookmarks/favourites and also cookies to be centrally stored so that they are available to the user from any terminal.

In the 'dynamic' section, there is information related to the user's current session or access mechanism as well as the ability to hold geographic location.

There are a number of commercial and standards-related activities taking place to define and exploit the profile opportunities, e.g. the Liberty Alliance [14]. Some of these are addressing the significance of the profile in managing identity. This leads

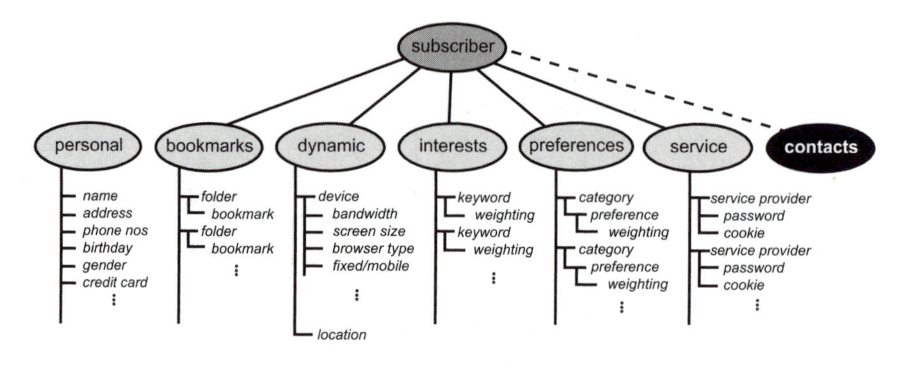

Fig 1.3 Personal user profile.

on to the significance of the profile for a range of identification functions [15] that are best addressed outside this chapter but include:

- identity — with varying degrees of anonymity;
- authentication;
- single sign-on.

For the profile to be valuable it relies on applications and services to populate it. Customisation can be a useful tool in this process as users will configure applications to their particular preferences and the profile can record this configuration data for subsequent analysis. Pooling personal interests and preferences can enhance the accuracy of this data, as it no longer relies on interaction with a single service provider.

1.6 Scenario — James the Architect

The following scenario is used to illustrate some potential applications and benefits of personalisation in an everyday context. It is also intended to highlight how applications can be integrated around a central personal profile and can generate revenue through charging for content and communications capabilities.

James is a self-employed architect and keeping in touch is essential to him for many reasons — customers, industry news, and promotion of his business. Over the last few years business has been going so well that he has employd a number of associates. This means that they now have to work very closely as a team on some complex projects, but they are forever on the road, on customer sites and in and out of meetings. Over the last year they have become dependent on mobile communications and the Internet to keep their business going.

One thing that James and his colleagues cannot cope without is their personalised information and communications services. Linking applications to a personal profile allows them to match their own interests, context and delivery medium. Following James through a typical day illustrates how the technology is applied and used.

It is a busy day today, so James gets up really early. Over breakfast he likes to catch up with the news. He does not bother with using his PC in his study much. He can keep his PDA with him all the time and high-speed networks keep him connected to the Internet.

A great thing about his portal is that he does not have to go sifting through the newspapers and architectural journals any more. His portal is set up to search for information that is important to him, and will even put the most important news where he sees it instantly.

Intelligent agents search for material of interest to James using the preferences he has set in his profile. The most likely ones are summarised using intelligent text summarising capabilities and presented to James in a prioritised list (see Fig 1.4).

Fig 1.4 Intelligent search agents.

Today's top news item concerns new legislation on building regulations. He quickly scans the headline and has the option to purchase a relevant report. He previews it with an automatic summary before he commits to the purchase. Once the payment is confirmed, his details are automatically placed in the rights management system to ensure that he has access to the content.

It dawns on him that this new legislation will affect his associates working on a town house project. It is essential that he brief them during the day. Fortunately he can submit the article to Jasper, the knowledge-sharing application, so that the right people will be alerted. He checks the town house project interest group listing to make sure that everyone is included (see Fig 1.5).

This morning, James needs to travel from Ipswich to Ely to visit a client who wants to refurbish their grade 2 listed cottage. The appointment is at 11 am and his

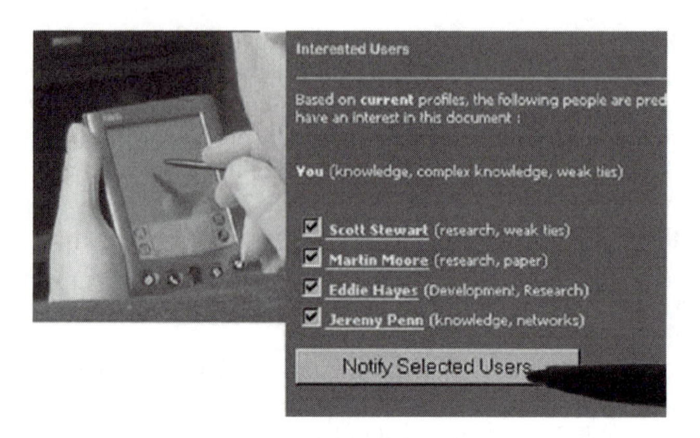

Fig 1.5 Interest groups.

PDA beeps to remind him to set off from Ipswich at 9.30 am. This allows him some extra time in case of delays. Of course, he had already entered the starting point and destination for his journey into the system. Putting his PDA in his briefcase James gets into the car and places his telephone in the docking station. He turns on the radio. He has set his profile to update him verbally on his progress every 10 minutes and to alert him if there are any traffic problems. Figure 1.6 illustrates how a personal profile system interacts.

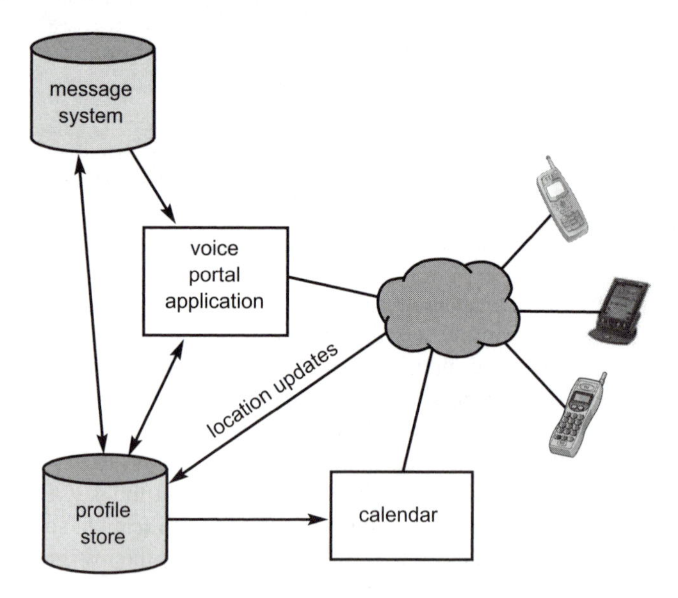

Fig 1.6 System interaction using personal profiles.

The telephone rings and the message system says:

"You are 42 miles from your destination and your approximate journey time will be 45 minutes."

The telephone rings again a few seconds later.

"E-mail message received from Karen Barker. Would you like to hear it?"

James: "Yes"

"E-mail reads:

'James. We need to talk about how the re-wiring will be done in the cottage refurbishment. Please give me a ring. Karen.'

Do you wish to reply to this message?"

James: "No, call her."

Telephone: "Connecting," followed by ring tones.

Telephone is answered. James: "Karen, hi, it's James. I'll be in the area around lunchtime."

This illustrates the power of sharing information across applications, and the power of multi-modal interfaces. James had configured his travel plans the day before and this information had been entered into the calendar section of his profile. His messaging profile was set to forward messages of high importance, including those where his geographic position was relevant. So, a message from a client is pretty important, but being in the client's local area made it even more important. The voice interface allows the user to retrieve textual information without the need for a keyboard or screen. Because the systems understand the concept that a single person has multiple contact points — phone, fax, e-mail, etc — a unified interface can be provided, enabling James to say, 'Call her'.

It's evening now and James has managed to get home relatively early. He's relaxing on the sofa, listening to some music from his personal music library. He finds it hard to remember that he used to use black vinyl, or even CDs... His portal gives him access to the music that he enjoys, wherever he happens to be. The best things are the reviews and recommendations. He even looks at the videos, which he never used to do. Then he remembers a new release that he heard on the radio earlier in the day and quickly pulls up the station's play list. He listens to a short sample from the album and decides to purchase it.

This application reuses the same payment and rights management services to allow James to listen to samples of the music, download the content and enable the licence key for access to it. This time he uses his own credit card for the payment.

1.7 Summary

Personalisation has often been dismissed due to its lack of success in certain applications. However, it is a much wider subject than is often assumed. The mobile telephone example shows how effective the concept of personalisation can be and how important it is to the way customers behave when selecting and using products and services. In particular, the success factors include:

- storage of personal contact data;
- convenience;
- personal communication;
- simplified billing, e.g. pre-pay.

The personal profile, when combined with Web Services technology provides an opportunity to integrate many applications and services together to provide cohesive and valuable propositions. Provided that privacy issues are properly tackled and users trust the host service providers, applications can interact in more intelligent ways so that the user is presented with enticing new offers.

The development of more sophisticated analysis techniques, in conjunction with the wider pool of profile data, will enhance the value of knowledge-management tools so that information is accessible in a more efficient way and can be shared more easily.

References

1 Oftel: '*Consumers' use of mobile telephony*', Summary of Oftel residential survey, Q7, November 2001 (January 2002) — http://www.oftel.gov.uk/

2 Zipf, G. K.: '*Psycho-biology of Languages*', MIT Press (1965).

3 Berry, M. J. A. and Linoff, G.: '*Mastering Data Mining: The Art and Science of Customer Relationship Management*', John Wiley and Sons (1999).

4 Shardanand, U. and Maes, P.: '*Social information filtering: algorithms for automating Word of Mouth*', Proc of the CHI-95 Conf, Denver, ACM Press (May 1995).

5 Mainspring Consulting Group (June 1999) — http://www.mainspring.com/

6 Jupiter (July 2000) — http://www.jup.com/

7 Peppers, D. and Rogers, M.: '*The one-to-one future*', Piatkus Books (1996).

8 W3C Web services homepage — http://www.w3.org/2002/ws/

9 Maes, P.: '*Agents that reduce work and information overload*', Comms of ACM, **37**(7), pp 31-40, ACM Press (July 1994).

10 http://www.friendsreunited.com/
http://groups.yahoo.com/

11 Hansen, E.: '*DoubleClick postpones data-merging plan*', CNET News.com (March 2000) — http://news.com.com/

12 Westin, A. F.: '*Consumer privacy in 2010: Ten Predictions*', Balancing Personalization and Privacy, San Francisco (June 2001).

13 Platform for Privacy Preferences (P3P) Initiative — http://www.w3.org/P3P/

14 Liberty Alliance — http://www.projectliberty.org/

15 Abelson, H. and Lessig, L.: '*Digital identity in cyberspace*', White paper submitted for 6.805/Law of Cyberspace: Social Protocols (December 1998).

2

AN OVERVIEW OF LOCATION-BASED SERVICES

T D'Roza and G Bilchev

2.1 Introduction

A location-based service (LBS) can be described as an application that is dependent on a certain location. Two broad categories of LBS can be defined as triggered and user-requested.

In a user-requested scenario, the user is retrieving the position once and uses it on subsequent requests for location-dependent information. This type of service usually involves either personal location (i.e. finding where you are) or services location (i.e. where is the nearest ...). Examples of this type of LBS are navigation (usually involving a map) and direction (routing information).

A triggered LBS by contrast relies on a condition set up in advance that, once fulfilled, retrieves the position of a given device. An example is when the user passes across the boundaries of the cells in a mobile network. Another example is in emergency services, where the call to the emergency centre triggers an automatic location request from the mobile network.

2.2 Positioning Technology

2.2.1 Global Positioning System (GPS)

Design of the global positioning system was begun in 1978 by the United States Department of Defense. Its original intended use was for military positioning, navigation and weapons aiming, but in 1984 following the crash of a civilian Korean Airlines flight the previous year due to poor navigational equipment, President Reagan announced that some of the capabilities of GPS would be made available for civilian use. In April 1995 the complete system containing 24 operational satellites in an 11 000 nautical mile orbit at a cost of $12bn was declared fully operational. In the early days of civilian GPS, the signal was intentionally scrambled through a process known as 'selective availability' — degrading the accuracy to around 100 metres.

Only the US-DoD benefited from the true accuracy of which GPS was capable; but at midnight on 1 May 2000 President Clinton ordered that selective availability be turned off [1], enabling typically 3—15 m accuracy for everyone. This decision was taken to encourage the already growing commercial, recreational and safety applications of GPS. Also, the possibility of GPS being used in a malicious way against the United States was no longer seen as a serious threat and in any case the US Military had demonstrated the ability to implement selective availability (SA) on a regional basis if required. The effect of SA is shown in Fig 2.1. Readings were taken from a standard GPS receiver at 30 sec intervals over a period of 24 hr at a continuously operating reference station in Kentucky [2] on 1 May 2000 and 3 May 2000 with and without SA respectively. Fig 2.1(a) shows that over time the position appears to 'wander' with 95% of readings falling within a radius of 45.0 m of the receiver's true location. Without SA, as shown in Fig 2.1(b), 95% of the points fall within a radius of 6.3 m.

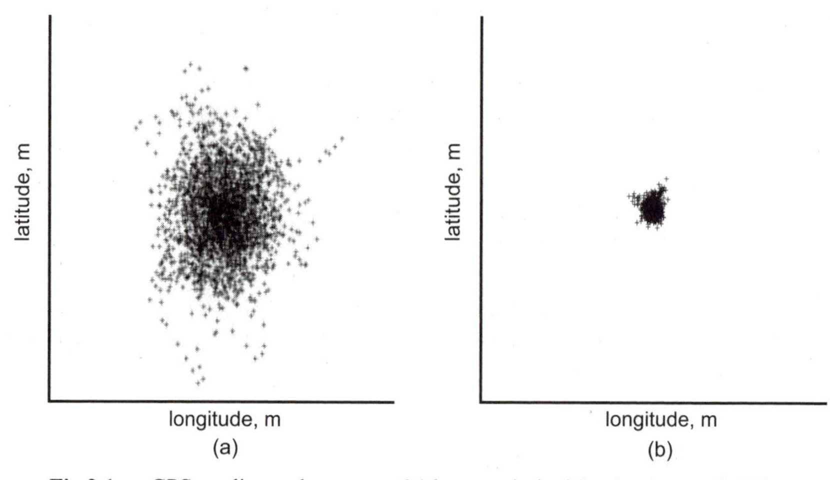

Fig 2.1 GPS readings taken over a 24-hour period with selective availability (a) turned on and (b) turned off [2].

The basic principle behind GPS is that a receiver measures the travel time of a pseudo-random code sent from the GPS satellite to the receiver — in practice, around 0.1 sec. From this the receiver can compute the distance (x) to the satellite which places the receiver somewhere on the surface of an imaginary sphere centred on the satellite with radius x (Fig 2.2(a)). The distance to a second satellite is then measured, narrowing the potential locations of the receiver to an elliptical ring at the intersection of the two spheres (Fig 2.2(b)). The potential locations of the receiver can be reduced further to just two possible points by incorporating measurements from a third satellite (Fig 2.2(c)). One of these positions is then disregarded due to it being either too far from the Earth's surface, or moving at unrealistic velocity. By taking readings from a fourth satellite, the receiver can be positioned in three-dimensions — latitude, longitude and altitude. The diagrams in Fig 2.2 show only a simplified 2-dimensional model.

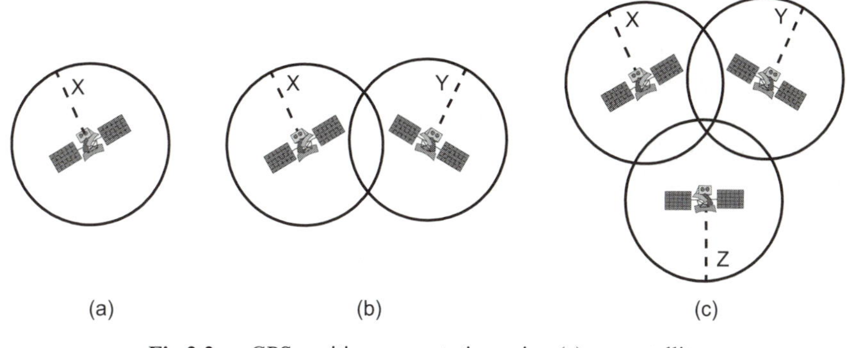

Fig 2.2 GPS position computation using (a) one satellite, (b) two satellites, (c) three satellites.

Although the basic principle is relatively simple, to achieve the desired accuracy requires the satellites to maintain precise orbits, which are continually monitored and corrected. Also, each GPS satellite contains four atomic clocks [3] synchronised to a universal time standard. Timing information as well as ephemeris updates — slight corrections to the satellite position table stored in every GPS receiver — are encoded in the satellite's signal. A secondary benefit is that the clock in a GPS receiver operates to atomic clock accuracy but at a fraction of the cost of a real atomic clock.

In the maritime environment, GPS is used as the primary source of positional data and is heavily relied upon for emergency and distress signals. By contrast, the aviation industry is not dependent upon GPS for its safety critical systems, only using it as an auxiliary location source.

2.2.1.1 Differential GPS (DGPS)

For most applications the accuracy of standard GPS is ample, but there are some applications (for example aircraft landing and manoeuvring boats in harbours) that require even greater accuracy. To satisfy this requirement, differential GPS was developed. Differential GPS provides a correction for errors that may have occurred in the satellite signal due to slight delays as the signal passes through the ionosphere and troposphere, and multipath errors. Differential GPS comprises a network of land-based reference stations at fixed locations on heavily surveyed sites where a very accurate position can be determined by means other than GPS. At these reference stations, GPS receivers calculate the position of the site, which is then compared to the actual known position and used to compute an error correction factor for each satellite. Because the GPS satellites are in such a high orbit, any mobile GPS receivers will be using signals that have travelled through virtually the same section of the atmosphere and so contain virtually the same errors as the signals received by the nearest reference station. GPS receivers can therefore apply the same correction factors that were computed by the reference station. Each reference station broadcasts correction factors to differential GPS receivers (e.g. 'add 15 ns to the signal from Satellite 01, 12 ns to Satellite 02', etc) on a separate radio network. Increasingly organisations that require this extra accuracy (such as the USA Coast Guard) are setting up their own differential reference stations and broadcasting the correction information free for public use.

2.2.1.2 Drawbacks of GPS

GPS offers unprecedented positional accuracy but it does have some drawbacks. Because the satellites are in a high orbit, and broadcasting over a large area the signal is very weak. The pseudo-random nature of the signal allows for small footprint antennas, but, because of the weak signal, the receiver needs a reasonably unobstructed view of the sky. This results in GPS receivers being unable to obtain a position fix inside buildings, under the cover of trees, or even when between tall buildings which restrict the view of the sky — an effect known as the 'urban canyon'.

2.2.2 GSM Cellular Location

Due to the cellular nature of the GSM mobile telephone network, it is possible to determine the location of a regular GSM mobile telephone. The basic system of cellID, described below, is somewhat crude but techniques are available to provide increased accuracy. This section describes one method of increasing the accuracy of cellID, but others also exist. The advantage of cellular positioning over GPS is that the signal is much stronger and therefore will operate indoors; it is also unaffected by the urban canyon effect (subject to GSM coverage).

2.2.2.1 CellID

CellID is the most basic form of cellular location and works simply by detecting the base transceiver station (BTS) with which the telephone is registered. At any moment in time the mobile telephone/station (MS) is registered to a BTS — this is usually the nearest BTS, but may occasionally be the BTS of a neighbouring cell due to terrain, cell overlap or if the nearest BTS is congested. Cells vary in size depending on terrain and the anticipated number of users — hence in city centres cells are much smaller than in rural locations (Fig 2.3(a)). This difference in cell size greatly affects the accuracy of a position fix since the location reported is in fact the location of the BTS and the MS may be anywhere within the boundary of the cell.

Typically the extent of error in urban locations may be around 500 m, but in rural locations this can increase up to about 15 km.

Each base-station will have multiple antennas, each covering a sector of the cell. So a BTS with three antennas will produce a cell with three 120° sectors. By detecting the antenna with which the MS is registered, the location of the MS can be narrowed down to somewhere within a sector of the cell with the BTS at its apex (Fig 2.3(b)).

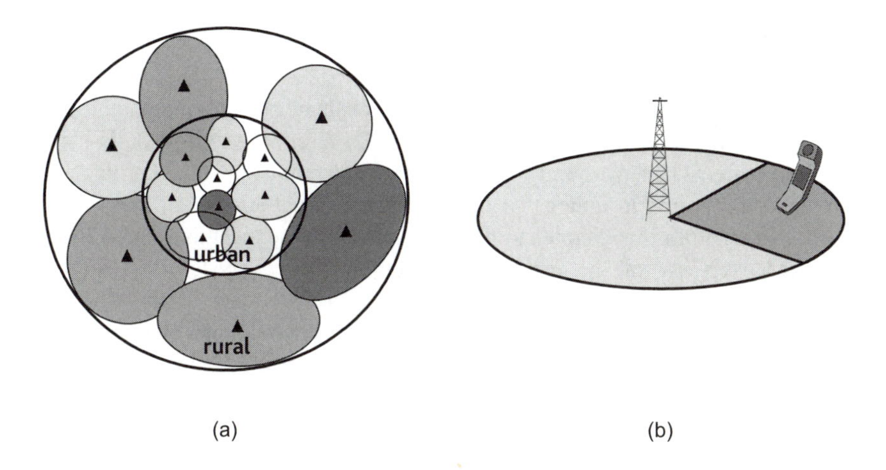

Fig 2.3 Base transceiver station size variation (a) and sectorisation (b).

2.2.2.2 *Enhanced-Observed Time Difference (E-OTD)*

E-OTD uses triangulation between BTSs to provide a more accurate location fix. The distance of the MS from the BTS is calculated by comparing the difference in time taken for a control signal sent by the BTS to arrive at the MS and at a fixed location — the location measurement unit (LMU) [4] — where the distance to the BTS is known (Fig 2.4(a)).

E-OTD can improve the accuracy of standard cellID by up to a factor of ten (Fig 2.4(b)). The disadvantage of E-OTD is that it requires significant investment by the network operator in installing the LMUs and a minor software upgrade to the handsets to enable them to calculate the signal delays.

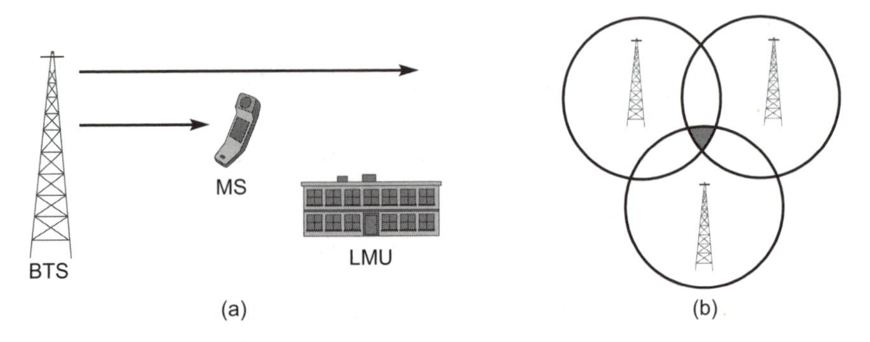

Fig 2.4 E-OTD's use of triangulation for a location fix is dependent upon LMUs (a) being available in adjacent cells (b).

2.3 Co-ordinate Systems

Today a number of different cartesian and polar co-ordinate systems exist, including universal transverse mercator (UTM), military grid reference system (MGRS), and national grid systems based around a range of reference points, units and projections used to represent points in either two or three-dimensional space. As there is not sufficient space within this chapter to discuss them all in detail, this section offers only a brief overview of the most widely adopted global system — latitude, longitude and altitude.

Latitude is a measurement of the angle at the Earth's centre, north or south of the equator. Longitude is a measurement of the angle at the Earth's centre, east or west of the prime meridian which runs pole to pole through Greenwich. Latitude and longitude can either be described in decimal format, or as degrees, minutes and seconds.

A good overview of co-ordinate systems is provided by Dana [5].

2.4 Representing Location in Applications

The ways in which positional information can be represented are largely dictated by the capabilities of the viewing device. Today the most powerful device tends to be a Web browser on a PC as it would normally have the greatest processing power, bandwidth, and screen size. A mobile telephone handset is perhaps the least powerful but there are plenty of other devices to consider including PDAs, multimedia kiosks, and digital TV. Regardless of the positioning method employed, the raw data returned is usually a co-ordinate, perhaps with an error parameter, for example '(52 03.50N, 001 16.89E) ± 5 m'. While this is by far the most accurate representation, to most users the raw data is of little use and only becomes valuable when interpreted in different ways.

A position can be described relative to a prior-known location in the same way that a person may describe 'The Red Lion pub' as being '100 m from the Post Office'. A table of geocoded points-of-interest is known as a gazetteer, and may typically contain town and city names, street names or business premises. For example, the co-ordinates in the example above would be described as '5.046 km South West of Woodbridge, England' when referenced against a 'towns and cities' gazetteer. A custom gazetteer may also be used in some applications to describe, for example, a company's own offices, depots and factories.

For many applications, including fleet management, tracking and routing, the most logical positional representation is graphical. There are many companies that specialise in the provision of map data including The AA [6], Ordnance Survey [7] and Bartholomew [8], with the main difference being one purely of aesthetics. A map image is usually comprised of a number of layers, where motorways, 'A' roads, city labels and county borders will all appear on separate layers. The application provider may choose which layers to include, and will usually opt to only include certain layers at particular zoom levels to avoid clutter.

There is one remaining factor that will influence the choice of map data — the choice between vector and raster data. As the name suggests, vector data (Fig 2.5(a)) is a set of instructions that describe the data in terms of vectors and so the resulting image will scale smoothly at different zoom levels. Raster (Fig 2.5(b)) data describes the map image pixel by pixel and so the image will appear pixelated when resized (Fig 2.5(c)). The choice is a trade-off between file size and map detail — vector data consumes less disk space and so is quicker to transfer to a remote device but will usually contain less detail than raster data.

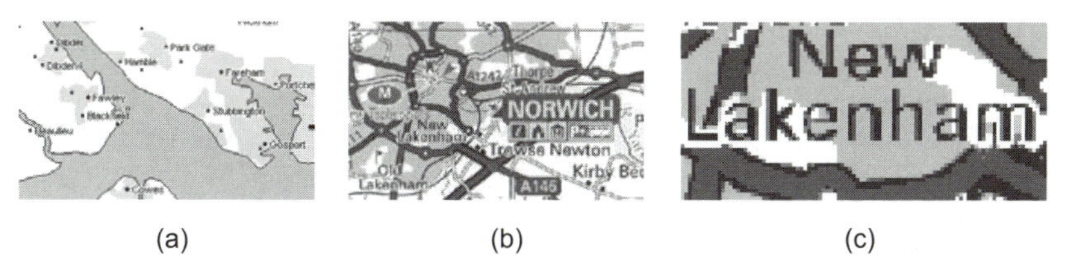

(a) (b) (c)

Fig 2.5 Map data as (a) vector, (b) raster and (c) pixelated raster after resizing.

2.5 Data Formats and Standards

2.5.1 XML

The geographic information system (GIS) industry has developed a set of XML description standard proposals. They address the two types of geography-related information — static (rivers, mountains, etc) and dynamic (events, moving objects). The former information is captured by the geography mark-up language (GML) and the latter is captured by the point-of-interest exchange (POIX) mark-up for the exchange of points of interest, the navigation mark-up language (NVML) for the mark-up of routes, and the SKiCAL for the mark-up of event information.

- GML

  GML [9] has been developed by the OpenGIS Consortium (OGC) to describe geographical features. It is part of a framework that is capable of digitally representing the Earth. GML is based on a common model of geography (OGC Abstract Specification) which is accepted industry-wide. It uses a hierarchically organised vector data model, which is well suited to the representation of geographic data. As expected, GML does not contain any styling information — this is up to the application. Usually an SVG (scalable vector graphics) can be used to transform GML into a representation (SVG is also XML based, so XSLT can be used to do the transformation). The drawbacks of GML include lack of format to store topology and no way to express co-ordinates with more than three dimensions (for example, time can be the fourth dimension).

- POIX

  POIX [10] is an XML-based language for the description of a position and information related only to that position. It was designed by the Mobile Information Standard Technical Committee. POIX is not capable of representing additional information such as opening times. It can, however, represent both a fixed or a moving position — the intent being to be able to describe a car as a point of interest as well as a restaurant. Point-of-interest descriptions are monolithic, but it is possible to link several points together.

- NVML

  The navigation mark-up language [11] is an XML format to represent locations as points along a route. NVML has been developed by Fujitsu as a data exchange format between navigation systems and to enable route data to be used in other contexts such as tourist information applications. NVML allows route descriptions to be generated dynamically thus adapting information to the path the user is actually taking.

- SKiCAL

  SKiCAL [12] is a special case of iCalendar VEVENT and is used to describe event information. iCalendar is used to create a format for vendor-independent exchange of calendaring and scheduling events. It enables users of different calendar software to book meetings and schedule events. SKiCAL extends the iCAL by providing a machine-understandable format for meta-information about events. It includes a vocabulary to describe the location, time and other relevant information about an event.

In summary, an application might use a combination of the above XML-based formats. This can be achieved by using XML namespaces [13]. It is also better to do the processing in the application server, which will guarantee sufficient capacity to handle generic XML processing.

Before moving on, one more point worth mentioning is the ability to search through XML mark-up. Although using standardised formats like GML, POIX, NVML and SKiCAL will ensure good results, it is even better to have a standardised way to embed meta-data information. One proposed format for that is the Dublin Core [14], which has been developed by the library community. The Dublin Core has elements for the description of author, publisher and other publication-related data, but it also includes an element for geographic and time-related metadata.

2.5.2 Oracle

Oracle Spatial [15] is a component of Oracle9i that comprises a set of functions and procedures to enable the storage of spatial information and allow it to be queried and analysed quickly and efficiently in an Oracle9i database. Spatial data represents the location of objects and how those objects relate to each other and to the space in which they exist.

Oracle Spatial is made up of the following components:

- a schema that prescribes the storage, syntax, and semantics of supported geometric data types;
- a spatial indexing mechanism;
- a set of operators and functions for performing area-of-interest (e.g. find nearest ...) queries, spatial join queries, and other spatial analysis operations;
- administrative utilities.

The components and utilities provided with Oracle Spatial facilitate the manipulation and maintenance of geometries (the geometric representation of a shape or shapes in some predefined co-ordinate system) in an object-relational way, including:

- support for many geometry types, including arcs, circles, compound polygons, compound line strings, and optimised rectangles;
- ease-of-use in creating and maintaining indexes and in performing spatial queries;
- index maintenance by the Oracle9i database server;
- geometries modelled in a single row and single column;
- optimal performance.

2.6 Applications

Service providers hope that location services will stimulate demand for wireless data services. Location information may be used by an application provider to personalise the service, or to improve the user interface by reducing the need to interact with a small device while on the move [16]. This section aims to give a brief insight into a range of likely applications of location-based services.

2.6.1 Communication

Some LBS applications will be self-contained — a user device obtains a position using one of the methods described above, performs some processing and then presents the resulting data back to the user. Many other applications will require the position to be sent to a server either for display to other parties, processing, or referencing against additional content. Consumer applications will often use SMS text messaging because it is simple to use and familiar to most mobile users. The disadvantage of SMS is that it is limited to text-based data (although the impending multimedia messaging service (MMS) will allow still images, audio, and video to be transmitted). WAP may be considered as an alternative communications channel that provides more data capacity and reduces the end-to-end delay, although, as has been widely reported in the press, take-up has been disappointing and the user interface is often cumbersome. SMS is also rather expensive as a data carrier and so may not be cost effective for some applications, where position reports need to be transmitted at 5-min intervals throughout the day, for example. GPRS may be a more appropriate bearer for some applications as only the data transmitted will be charged for, and the high data rates would allow for large position and telemetry logs to be downloaded at the end of the day if required. All of the communications channels discussed so far have relied on the GSM network, but for safety-critical applications, or tracking of devices in remote areas, GSM may not be appropriate. A satellite network, such as Inmarsat C or D+, may be preferable if global coverage is required, although there will be an obvious trade-off with cost per-position-report, and the hardware is likely to be more bulky and demand more power.

2.6.2 Fleet Management

The purpose of a fleet management application is to allow a company to keep track of its mobile assets, in near-real time, and to be able to use that information not only to increase performance and utilisation but also decrease operating costs. As an example, consider the case of a delivery company.

By having its fleet of delivery vans reporting their position at regular intervals throughout the day, if an urgent collection is required, the company knows which is the nearest van and can calculate the travel time required, therefore optimising the distribution of tasks. If the vehicle is also reporting telemetry data about engine performance and driving habits (acceleration, braking, etc) the company can also detect mechanical problems before they cause damage, and encourage their drivers to adopt a more fuel efficient driving behaviour. Geographic boundaries, known as geofences, could be configured that trigger alerts when the object being tracked crosses the geofence perimeter. These could be defined so that when the lorry arrives within 5 miles of the depot an alert is triggered to forewarn the loading bay crew of the van's arrival. Location data could also be viewed by customers to better inform them about the location of their deliveries and expected delivery time.

2.6.3 Routing

Satellite navigation is another increasingly common implementation of location-based services and the benefits in terms of optimised routing, avoidance of traffic congestion and early warning of diversions, accidents and roadworks are easy to recognise. Users found early systems frustrating as they did not always take account of restrictions such as one-way systems and if incorrectly configured may have tried to direct the driver along the shortest route in terms of distance (which may have involved driving through the centre of a town at rush hour), rather than travel time. Modern systems have rectified many of the early complaints and often allow the map data to be updated and may also take account of traffic congestion and variations at different times of day.

Apart from detailed turn-by-turn directions, there is growing demand for 'Where's my nearest ...?' type applications where an end user requests the nearest business of a particular type relative to their current location, for example 'Where's my nearest Italian restaurant?' To date these applications have relied on self-positioning by the user where the user has to define their location manually either by entering a street name, town name, postcode or some other reference. This is because until now it has not been possible for a third party application provider to determine the location of a user's mobile handset; but the imminent roll-out of APIs to the network's CellID data will provide a significant boost to these services.

2.6.4 Safety and Security

An emerging application of location-based services is in the area of workforce safety. By equipping their workforce with a small electronic device that enables location determination and transmission into a service centre, a company can monitor the condition of lone workers and those in high-risk areas. Status updates may be requested at regular intervals, and the device may have a 'panic button' to allow the user to request that assistance be despatched to their precise location in the event of an emergency.

Vehicles can now be equipped with covertly installed tracking devices to allow their safe recovery in the event of theft. Many of these systems are so successful that motor insurance companies now offer discounts to the insurance premiums of those that choose to have the relevant devices installed.

2.6.5 Entertainment

The limited availability of low-cost, mass-market positioning devices has so far been a barrier to location-based services entering the entertainment arena — to date specialised GPS hardware has been required. However, the combination of the ever-decreasing price of GPS technology, and the imminent availability of GSM CellID positioning, has contributed to the appearance of some innovative entertainment applications. Location-based directory services have been around for some time using either a WAP or SMS interface. Community-based applications are now being launched within bars and night-clubs that allow messaging on a person-to-person basis within the same location. Example applications for this type of service are DJ requests, voting, competitions, and dating services. Many applications within the entertainment sector will be enhanced by the MMS applications currently being rolled out by GSM operators in the UK.

2.6.6 Business Models

This is a topic of utmost importance to the mobile operators (MOs) who paid billions of pounds for a third-generation licence that is yet to be used. Once the MO knows the position of the user, it is easy to get greedy and hence the idea to charge users for access to the positioning information. This does not, however, build trust. Therefore, this charging model should be applied with caution. There are a number of ways the users could get charged for an LBS:

- per request;
- by subscription;
- a combination of the two.

Mobile operators could also look at revenue-sharing deals, where the users are charged for an LBS (not just for position information) and the MO gets a cut of the generated revenues. The MOs can also become payment aggregators and providers by using the mobile telephone bill to charge the users for all mobile services.

A supplementary source of income can be location-based advertising. Because the mobile telephone is very personal, the solution might be to create even more personalised (to user profile and user current location) advertisements; for more details see Chapter 11. The marketing advantage of location-based advertisements is clear — immediately measurable success rate. Extra value-added features, such as coupons, can be included to attract users. Advertisers could be charged by the response rate or using the similar notion of click-through from the Web — the number of times information about the advertisers has been sought.

If a service provider has a sufficiently strong brand and its services are generating a lot of traffic volume, it might consider asking the MO to pay for generated increased traffic. This idea, however, is largely alien to MOs, especially if they charge users a subscription fee for access to services rather than per-volume of traffic. That said, however, the emergence of 'virtual operators' and the downturn in the economic climate might push operators to reconsider their business models.

2.6.7 Billing

Finally, this section considers billing mechanisms for LBS applications. The network operators take the position that they own the location information and hence could sell it to providers of LBS. This also allows them to provide the payment service and do the final billing to the users. This model is not very much different from the NTT DoCoMo model, where the MO charges 9 per cent of the transaction sum. This helps the service providers in at least two ways:

- users are not required to register with each service provider, because of the low value of each transaction and the age of the users,
- a credit card solution might not work.

The logging of chargeable events is not very difficult. The problem lies in getting that information into the system where the bill is created and sent to the user. Operators are often tied into their old big monolithic billing systems that would require considerable effort to be fully integrated.

Transactions are captured by customer data records (CDRs) and recorded into a database. This can either be in the billing database or separately. This can happen in real time (as in the prepaid mobile accounts) or in a batch mode (once a month for the post-paid mobile accounts). The logic of how much to charge for each CDR could reside either in the billing system or the application server. In the latter case the important point is that the interfaces are secure. How the interface works is dependent on the architecture, but, in principle, the application server should be able to create a record in the billing CDR database.

There is also a connection to the provisioning system, which controls access to the available services. This could, for example, prevent users from getting access to a service that requires prior subscription.

MOs (especially in Europe) have recently realised the value of becoming micro-payment service providers by offering a payment interface to service providers (through the use of reverse-charged SMS). Although not a generic micro-payment engine (i.e. only certain predefined charge bands are offered), reversed-charge SMS is proving to be a popular third-party interface into the MO's billing systems.

2.7 Summary

In the USA, the market for location-based services is being stimulated by the E911 mandate that requires telephony operators to be able to locate the source of emergency calls. Phase I effectively requires CellID to be in place, whereas Phase II requires far greater accuracy — within 50 to 100 metres in most cases [17]. The low investment required to implement CellID has made it an attractive value-add service for network operators resulting in the roll-out of a system in Canada without the need for a Government mandate. The E911 regulations have driven the development of positioning technology which will in turn drive LBS applications. In Europe, similar legislation (known as E112) is being proposed by the European Commission [18].

GSM network operators are increasingly eager to migrate their user-base from voice-centric applications to more profitable data applications in order to increase their average revenue per user (ARPU), thus helping to recoup their investment in third generation network licences and infrastructure. Hardware costs continue to fall while device capability increases, and though we may not yet have seen the 'killer application', it may be just around the corner — and location-based services could help us to find it.

References

1 Office of the Press Secretary, The White House — http://www.navcen.uscg.gov/news/

2 National Geodetic Survey — http://www.ngs.noaa.gov/FGCS/info/sans_SA/compare/ERLA.htm

3 '*Accuracy is addictive*', The Economist (2002) — http://www.economist.com/science/tq/displayStory.cfm?story_id= 1020779

4 Cambridge Positioning Systems — http://www.cursor-system.com/

5 Dana, P. H.: '*Coordinate systems overview*', — http://www.colorado.edu/geography/gcraft/notes/coordsys/coordsys_f.html

6 The Automobile Association — http://www.theaa.com

7 Ordnance Survey — http://www.ordnancesurvey.co.uk

8 Bartholomew — http://www.bartholomewmaps.com/

9 GML (Geography mark-up language) — http://opengis.net/gml/01-029/GML2.html

10 POIX (Point-of-interest exchange language) — http://www.w3.org/TR/poix/

11 NVML (Navigation mark-up language) — http://www.w3.org/TR/NVML.html

12 SKiCAL — http://www.globecom.net/ietf/draft/draft-many-ical-ski-03.html

13 XML namespaces — http://www.w3.org/TR/REC-xml-names/

14 Dublin Core — http://dublincore.org/

15 Oracle Spacial — http://download-west.oracle.com/otn/oracle9i/901_doc/appdev.901/a88805/sdo_intr.htm

16 Basso, M. and Kreizman, G.: '*Mobile location services for governments*', Gartner (September 2002).

17 Federal Communications Commission (FCC) 911 Services (August 2002) — http://www.fcc.gov/911/enhanced/

18 GI News E112: '*Europe states its position*', (September 2002) — http://www.ginews.co.uk/0902_18(1).html

3

LOCATING CALLS TO THE EMERGENCY SERVICES

P H Salmon

3.1 Introduction

Provision of emergency services varies widely between countries since they have evolved to meet national needs and have not had any good reason to harmonise with other countries. In some countries, for instance, fire service personnel also act as paramedics. Unsurprisingly, the emergency telephone numbers vary between countries. Some countries have one number for all services while others have separate numbers for each service and these can be different again in separate regions within a country. Often, sections of the population of a country are not aware of what number they should call in an emergency. A visitor to a foreign country is even less likely to know the correct numbers. The EU has selected 112 as the emergency services number for the EU countries. Concerns were expressed about using a low number as line noise can generate false calls of numbers with 1s and 2s in them. Though 112 is available in most European countries, there is little public awareness of it. A survey showed that many Europeans were more aware of the American 911 because of the TV programme *Emergency 911*. Once a call has been received by the emergency services it is important that help is despatched as soon as possible to the incident location.

3.2 The Importance of Location in Emergencies

In an emergency situation people are often confused and, in panic, unable to say where they are calling from. Time spent on determining the caller's location leads to delay in the despatch of the emergency services and can also block other incoming emergency calls. In addition, it increases stress on the operator who is trying to help the caller as efficiently as possible. Paramedics refer to the first hour after an injury as the golden hour where early treatment has a significant effect on survival rates.

Major incidents that are witnessed by many people usually lead to a surge in calls to the emergency services as people ring in on their mobiles. Apart from filtering on the location of mobiles there are other ways to mitigate this problem; for instance, in Sweden an automated announcement is played saying, for example, '... if your call is not about the large fire in central Stockholm, please press 4.' Motorways can be a particular problem as people ring in as they pass an accident and can be a few miles away when they actually make the call. A cluster of calls are produced and the emergency authorities are not sure if they have one or multiple accidents to deal with.

Calls made from inadvertent key presses while the mobile is in a pocket or bag lead to a large number of silent calls being received by the emergency operators. Up to 20% of emergency calls can be silent and often they are from mobiles with the key lock on [1]. Formerly, police control rooms would have had to decide whether to close calls but they are now routed to a call-handling system which asks the caller to press '5' twice if they require the emergency services. In mid-2002 over 400 000 silent calls per month were taken by the system.

Over recent years, small-scale implementations and trials of in-vehicle emergency calls have been operating. An emergency call, sometimes with a position obtained by a GPS receiver, is made when the telematics unit receives a trigger from the vehicle such as the air bags have inflated. The telematics unit is usually part of an overall package installed by the manufacturer that, among other functions, can automatically make a call about a breakdown or emergency. One issue with this service has been that the emergency call is sometimes routed to the manufacturer's call centre which then has to route it on to the relevant emergency services. There is an understandable desire from the manufacturers to own the customer, but this view has not been shared by some of the emergency services. Another problem that has arisen is that in an accident the occupant leaves the vehicle and makes an emergency call while the vehicle has also initiated a call. An accurate location associated with the call may reveal this type of incident. In future, it is likely that there will be an increased incidence of in-vehicle emergency call systems.

3.3 Public Safety Answering Point (PSAP)

Emergency call-handling centres are known as PSAPs. Emergency calls are routed by the network to PSAPs where they are answered by an operator. PSAPs are divided into stage 1 and stage 2 PSAPs (see Fig 3.1). In the UK, mobile calls are routed by appending the network telephone number, or calling line identity (CLI), and cellID (or a zone code corresponding to groups of cells) to the C7 call set-up message. The calls are answered at a stage 1 PSAP where an operator filters out false calls and routes the remaining calls using cell/zone to the appropriate local emergency services at the stage 2 PSAP. In most EU countries the emergency services run the stage 1 PSAPs though in the UK the stage 1 PSAPs are operated by

BT and Cable & Wireless who forward the CLI automatically to the stage 2 PSAP, again using the C7 set-up message. Stage 2 PSAPs are operated by the emergency services and are responsible for the despatch of the necessary assistance to the incident. Within the EU there are between 14 and 1060 stage 1 PSAPs in each country and numerous stage 2 PSAPs [2]. In the UK there are over 300 stage 2 PSAPs. The network is robustly designed to cope with the failure of intermediate links or switches.

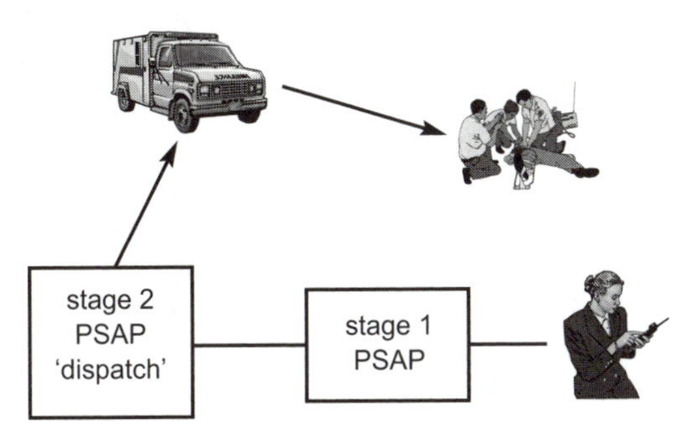

Fig 3.1 Public safety answering points operation.

Databases for the fixed network maintain the name and address of the customer. These are updated on a daily basis by the operators and can be accessed by the PSAPs, though the level and ease of access varies widely between countries. There are some problems in locating fixed telephones for those countries that do not have an automated method for PSAPs to access the installation address of the CLI. Calls made from switchboards (PABXs) for private networks do not always have the correct location of an extension user associated with them. In some instances a number of geographically separate sites could be routed through a single organisation's PABX, and only the organisation concerned knows where each extension is sited and does not make that information available to the network provider. In the UK these multi-site problems are handled using a warning flag on the address information obtained from databases. There is very little information available to the PSAPs about mobile callers. For prepaid callers even the operators often do not know the identity of the customer.

3.4 The US Federal Communications Commission (FCC) 911 Mandate

The impetus for accurately determining the location of a mobile has come from the US emergency authorities, who had become concerned about the time it was taking to resolve where the growing proportion of cellular emergency calls were located as over 50% of USA emergency calls are mobile originated. After some years of lobbying, the FCC ruled in 1996 that mobile operators would, by October 2001, automatically provide the PSAP operator with the location of the user according to the following accuracy:

- for network-based solutions — 100 metres for 67% of calls, 300 metres for 95% of calls;
- for handset-based solutions — 50 metres for 67% of calls, 150 metres for 95% of calls.

Since 1996 there has been clarification and revision of these rules. The majority of mobile users should be covered by automatic location information (ALI) by 2004. Technology development did not meet the time-scales and all the major operators applied for waivers from FCC to modify criteria such as the implementation time-scales and accuracy. Sprint and Verizon both launched a service on 1 October 2001. Some operators have been fined by the FCC as they were judged to have made insufficient effort to meet the targets.

In the USA, a mobile telephone bill has a separate line item for the provision and maintenance of the emergency operator service and this varies state by state, whereas in Europe this is invisible to the customer and absorbed by the operator. Also in the USA, the increased costs of providing location information will be explicitly billed to the end user.

3.5 European Emergency Location Initiatives

The Universal Services Directive [3] is part of the new regulatory framework for electronic communications networks and services that should be implemented across the EU by July 2003. This best-effort approach by the operators and the emergency services will be reviewed by 2004 to see if any additional actions are required. An extract of article 36 that applies to caller location for both fixed and mobile telephones is given below:

> 'Caller location information, to be made available to the emergency services, will improve the level of protection and the security of users of '112' services and assist the emergency services, to the extent technically feasible, in the discharge of their duties, provided that the transfer of calls and associated data to the emergency services concerned is guaranteed. The reception and use of such information should comply with relevant Community law on the processing of personal data. Steady information technology improvements will progressively support the simultaneous handling of several languages over the networks at a reasonable cost. This in turn will ensure additional safety for European citizens using the '112' emergency call number.'

Civil protection is an area of national responsibility rather than EU, but when combined with telecommunications there is some overlap.

3.6 Co-ordination Group on Access to Location Information by Emergency Services (CGALIES)

CGALIES was established by the EU in 1999. The terms of reference evolved as CGALIES developed and the mission and mandate from February 2001 [4] is given below:

> 'The main task of CGALIES is to identify the relevant implementation issues with regard to enhancing emergency services in Europe with the provision of location information, to analyse and describe them, and to build a consensus on the Europe-wide implementation, involving the views and opinions of all relevant players. In certain cases, issues are perhaps better resolved at a national level. Notwithstanding this, it might be important and appropriate to discuss them at the European level to facilitate the consensus-building process.'

To complete the work of CGALIES, three working groups (WGs) were created consisting of operators, manufacturers and emergency service representatives.

3.6.1 WG1 — Location Technologies

This group was responsible for obtaining the emergency service requirements for location accuracy and reliability, and the time-table for these requirements being achieved using available technologies.

3.6.2 WG2 — Public Safety Answering Points

The responsibilities of this group were to draw together the differing emergency service models from across Europe in order to make recommendations on common interfaces and standards for PSAPs.

3.6.3 WG3 — Financing and Cost Analysis

This group studied the cost of providing location information and how these costs might be financed.

3.7 European Emergency Service Requirements

Separate stages of the emergency response process demand different accuracy resolution. When an emergency call is made, the initial call routing to the correct stage 1 PSAP does not require precise accuracy — 1–10 km accuracy in urban/ suburban areas and 35 km in rural areas. The despatching of the appropriate emergency personnel requires greater accuracy — 500 m in urban areas, 5 km in suburban areas and 35 km in rural areas. The requirements for caller location are given in Table 3.1.

Table 3.1 Emergency services caller location requirements.

Caller Status	Rural	Suburban	Urban	Motorways/ Waterways
Caller is able to provide location information and a location estimate is available	100-50 m	50-50 m	25-150 m	100-500 m
Caller is unable to provide location information and a location estimate is available	10-500 m	10-50 m	10-15 m	10-50 m

3.8 Mobile Location Technology for Emergency Services

The horizontal accuracy performance of current location technologies for single measurements is shown in Table 3.2. Accuracy for all these technologies, except cellID, is improved by making multiple measurements. CellID is the technique of locating a user according to the base-station with which their mobile has established a session. The operator is able to look up the identity of the base-station to find its geographical location. The table is updated regularly because network upgrades often reassign identities between base-stations. CellID and timing advance (TA) determines the position to within a ring from the base-station. Timing advance compensates for the differences in propagation delay experienced by mobiles as their distance varies from the base-station. Enhanced cell global identity (ECGI) improves the accuracy of cellID and timing advance by using the measurement reports of field strength from surrounding base-stations. Enhanced observed time difference (EOTD) compares the arrival time of network signals at the mobile with the arrival time at network measuring devices, and inputs these measurements into a triangulation calculation. At least three base-stations are required to make an EOTD measurement (see Fig 3.2). With assisted global positioning service (AGPS), the network provides GPS [6] data to the mobile to expedite the calculation of its position from the GPS signals it can receive. In general, EOTD has better performance in urban and suburban areas than in rural areas. Conversely, AGPS has its best performance in the rural areas and its worst in urban areas and indoors where the satellite signals are heavily attenuated. It is not possible to be categoric about performance because of the nature of the radio environment. The location estimates produced by these methods always have an associated error bound.

Table 3.2 Horizontal accuracy of location technologies [5].

Technology	Rural	Rural extreme	Suburban	Suburban extreme	Urban	Urban extreme	Indoor user	Comments
CellID	1-35 km	1-100 km	1-10 km	1-10 km	50 m-1 km	50 m-1 km	No change unless there is a pico-cell	Cell shape can be returned, possibility of incorrect sector
CellID and timing advance	1-35 km	1-100 km	1-10 km	1-10 km	50 m-1 km	50 m-1 km	No change unless there is a pico-cell	Radial distance can be improved for ranges above 550 m, possibility of incorrect sector
ECGI	250 m-8 km		250 m-2.5 km		50-550 m		50-550 m	More accurate than cellID + TA
EOTD	50-150 m	50-150 m or unavailable if not 3 basestations	50-150 m	100-250 m	50-150 m	100-300 m	Slight degradation but penetrates well indoors	Mobile needs to see at least 3 base stations, falls back to cellID/TA if unavailable
AGPS (to be confirmed)	10 m	10 m	20 m	50-100 m	30-100 m	50-100 m if available	In-building coverage by windows but not deep inside	CellID fall-back

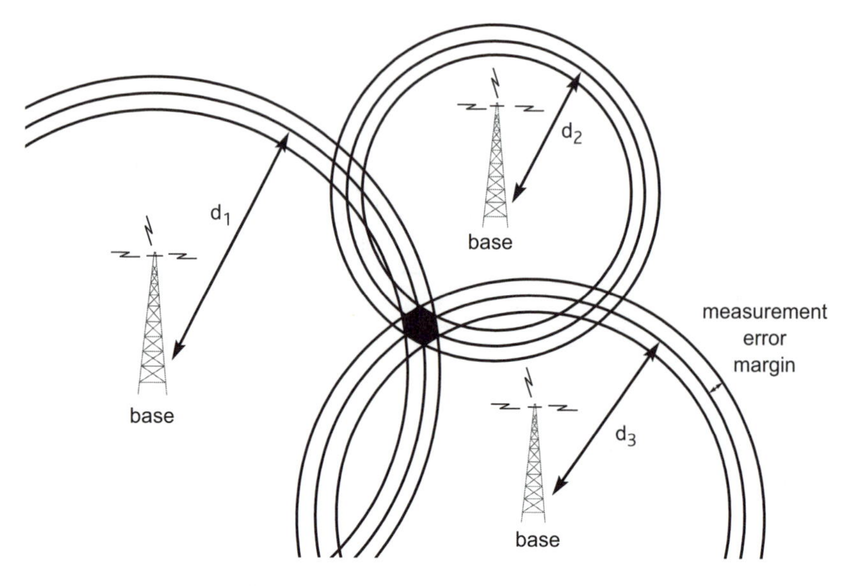

Fig 3.2 EOTD position calaculation.

There are numerous cellID-based location services from operators. CellID has been deployed in Estonia for the emergency services in such a way that the estimated location of the user is displayed on the emergency despatchers' screens. Interestingly, the initial feedback from the emergency services of Estonia has been that the location has proved very useful in rural areas but they would benefit from greater accuracy in urban areas, though urban cells cover smaller areas than rural and give a better location estimate. The early hypothesis is that the paucity of the road network in rural areas compensates for the more inaccurate location.

It is going to take time for these technologies to be built into devices by manufacturers. Users tend to replace their handsets every 2—3 years and from this the location technology penetration can be estimated. A reasonable definition of complete penetration is around 97% to take into account people that have bought handsets in overseas markets and roamers. Some manufacturers are already building an EOTD capability into their handsets as the cost to them of doing this is negligible. This is analogous to the development of SMS where handsets were SMS capable long before the networks supported SMS. Generally speaking, by the time the operator supports the capability, there is already a large user base that could potentially use the service. The network infrastructure needed to provide location services will also take the operator many months to procure and deploy across the network. Sprint have already sold 700 000 AGPS handsets in the USA [7].

The latency from when a location estimate is requested to when it is calculated is crucial. It is acceptable for a coarse estimate to be provided quickly to allow the call to be routed to the stage 1 PSAP, which generally covers a large area, as long as a more accurate position can be given that will allow the call to be onward routed to the relevant stage 2 PSAP. The accurate location estimate should be performed in 15—30 sec to be useful. The capability to provide speed and direction can be beneficial in cases such as motorway accidents where it would assist the emergency services in determining whether multiple callers are describing single or multiple accidents.

Borders are a problem because radio coverage plans do not map perfectly to political boundaries. An emergency call made in one country may be received by a base-station in another country, depending upon the caller's rights to roam on to other networks. Neighbouring countries have established procedures to handle these calls though mostly through manual intervention. An accurate location estimate could diminish the number of border incidents and promote automated forwarding in the event of a caller being incorrectly located across a national boundary. Mobile operators could select different location technologies from mobile operators in other countries. A roaming user could find that a visited country does not support the location technology of their home operator. The quality of their location estimate would be reduced and the location estimate would default to cellID.

Though absolute horizontal accuracy is the essential factor in any location technology there are many other important criteria. For any high-rise building, the ability to know from which floor a call is made can reduce the amount of searching that the emergency services need to do. The available location technologies do not provide sufficient vertical accuracy to be useful. Pico-cells may have a useful role to play in providing floor level detail, but are not yet installed in large enough quantities or adequately standardised to be useful.

The consistency of any location service needs to be understood. The availability of location estimates can vary according to the time of the day. The number of GPS satellites that are visible from a particular point on earth varies as they elliptically orbit the earth [6]. The USA reserves the right to change the GPS orbits for military reasons. Certain areas of a mobile operator's radio coverage will have greater accuracy location coverage than other areas. In some areas of radio coverage it may not be possible to get a location estimate better than cellID. EOTD requires reception from at least three base-stations to yield a location estimate. As long as there is radio coverage it will always be possible to provide cellID.

3.9 Standards

The first location standards were completed in 2000 by the USA-based T1P1.5 LCS (location services) group. T1P1.5 LCS had taken the lead in developing GSM location standards in order to meet the FCC 911 mandate. Location standards became part of GSM release 98 and UMTS release 99. Subsequent GSM and UMTS releases have further specified the emergency behaviour of LCS.

There has been a proliferation of organisations concerned with emergency telecommunications. Among others, the United Nations Working Group on Emergency Telecommunications met in Rome during June 2002.

During June 2002 ETSI [8] created the EMTEL Ad Hoc Group to co-ordinate emergency telecommunications requirements across ETSI deliverables. The group intends to restrict itself to producing requirements, while the other relevant ETSI groups produce the standards that meet the emergency telecommunications requirements.

Since Spring 2002, an Oftel Public Network Operators Interoperability Interconnection Standards Committee Task Group has been meeting to develop a common specification for an interface for the transfer of emergency location information from mobile networks to the stage 1 PSAPs. The membership comprises the mobile operators, the stage 1 PSAPs and the manufacturers. The interface is derived from the Location Interoperability Forum (LIF) mobile location protocol (MLP) [9] using XML/HTTP. Initially, a PULL service from the emergency operator will be provided before the mobile operators provide an automated PUSH service. The mobile operators will supply the terminal's mobile station ISDN number (MSISDN), the grid reference of the caller, the area in which the caller is situated (within a confidence level), the date, and the time when the location estimate is made.

3.10 Location Privacy

It is assumed that a user has waived their right to anonymity of location by making an emergency call. Concerns have been expressed that revealing a caller's location may deter some from calling the emergency services though it is thought that it is more useful to the emergency services to know where a caller is, than that the person will not call. For all other location services the EU user has the right to anonymity and even when they have subscribed to a location service they have the right to withhold their location at any time. The EU citizen owns their location. At the time of writing, the Data Protection Directive was in draft form, but it is unlikely that any of the articles concerning location and privacy will change significantly. In the USA, the situation is different and it is the operator that owns the user's location.

3.11 Geographic Information Systems

To make optimal use of the location estimates that are provided concurrently with the emergency call, the emergency despatcher's workstation requires the ability to display the caller's location automatically. Typically, digital maps are accurate to about 10 m and the estimate of the caller's position should be overlaid on to the map, perhaps with any other information that is available about the incoming number such as the name and address to which the number is registered. The location error estimate is usually displayed as an ellipse with a confidence level of 67% or 95%.

3.12 Summary

The effective use of location to assist the emergency services requires the co-operation of the telecommunications operators, manufacturers and the emergency services. These organisations are working together to provide an enhanced service. Co-ordination between these groups is essential to enable the various components to combine at the appropriate time without undue costs. Mobile operators are concerned that they may be required to provide a level of accuracy that is not commercially justifiable. Some emergency service representatives argue that it would be better to buy more ambulances, for example, than to upgrade PSAPs to the standards necessary for the use of location data. The preferred route to optimise benefit is for the mobile operators to provide their location information as it becomes commercially available as part of their deployment of location-based services. From July 2003 UK mobile operators have been using a standard interface to provide location information to the PSAPs.

References

1 '*The Silencing of the CWSR (or how the Metropolitan Police have quietly dealt with the silent 999 menace)*', The BAPCO Journal (July/August 2002).

2 Malenstein, J., Ludden, B., Pickford, A., Medland, J., Johnson, H., Brandon, F., Axelsson, L. E., Viddal-Ervik, K., Dorgelo, B. and Boroski, E.: '*CGALIES — final report on implementation issues related to access to location information by emergency services (E112) in the European Union*', (January 2002) — http://www.telematica.de/cgalies/

3 Directive 2002/22/EC of the European Parliament and of the Council: '*On universal service and users' rights relating to electronic communications networks and services (Universal Service Directive)*', Official Journal of the European Communities (March 2002).

4 Co-ordination Group on Access to Location Information by Emergency Services (CGALIES): '*Terms of Reference*', (February 2001) — http://www.telematica.de/cgalies/

5 Ludden, B., Malenstein, J., Pickford, A., Vincent, J. P., Heikkinen, P., Paris, G., Salmon, P., Paul, R., Dunn, T., Davies, R. and Evans, S.: '*Co-ordination Group on Access to Location Information by Emergency Services Work, Package 1*', (May 2001) — http://www.telematica.de/cgalies/

6 Hofmann-Wellenhof, B., Lichtenegger, H. and Collins, J.: '*GPS Theory and Practice*', 4th edition, Springer Verlag (1997).

7 pulver.com LBS report (July 2002) — http://wwww.pulver.com/

8 ETSI — http://www.etsi.org/

9 Location Interoperability Forum — http://www.locationforum.org/

4

LOCATION-BASED SERVICES — AN OVERVIEW OF THE STANDARDS

P M Adams, G W B Ashwell and R Baxter

4.1 Introduction

Interest in location-based service (LBS) standards began in the 1990s with the more advanced second generation cellular systems — particularly in GSM. There has been a general drive from mobile operators to increase their average revenue per user (ARPU), and by the late 1990s it was clear that the cellular markets were fast approaching the saturation point in terms of customer penetration rates. Therefore the only way to further increase the turnover of mobile operators was to raise their ARPU by the successful provision of more value-added services, such as the LBS.

Mobile systems, such as GSM, have always been heavily driven by standards so as to achieve full interoperability between different suppliers as well as international roaming on to foreign networks[1]. The focus for GSM standards developments throughout the 1990s was the European Telecommunications Standards Institute (ETSI) Special Mobile Group (SMG). However, for certain specialist functions, such as the development of end-to-end protocol solutions between mobiles, ETSI SMG has worked with other groups such as the Wireless Application Protocol (WAP) Forum.

4.2 Evolution of the Framework for the Development of LBS Standards

4.2.1 The North American Influence on LBS Developments

In the mid-to-late 1990s an increasing number of American cellular operators started to adopt GSM in their home networks. GSM had an attractive range of

[1] Generally roaming on to other mobile networks in one's home country has not been possible in 2G systems, but this was one of the policy issues being re-evaluated for 3G systems like UMTS.

features which, when combined with an aggressive export marketing campaign by multi-vendor suppliers, such as Nokia, Ericsson, Alcatel and Nortel, led to a breakthrough for the European cellular industry in the North American market. These American cellular operators became known as 'GSM North America' and the list of operators joining them has become progressively longer and now includes such well known North American names as Vodafone Airtouch and Cingular Wireless Systems. For LBS, the importance of the move of GSM into the North American market was the existence of stringent FCC regulations on location requirements for terminals making emergency calls, which for 3G can be summarised as follows:

- in 67% of cases mobiles must be able to be located to the nearest 50 metres for emergency calls using handset-based positioning methods;
- in 67% of cases mobiles must be able to be located to the nearest 100 metres for emergency calls using network-based positioning methods;
- emergency services do not currently require information on the vertical location of the user.

These requirements are hard to meet technically — particularly using the W-CDMA technology chosen for UMTS. However, at the IEE 3G Mobile Conference in London in May 2002, key suppliers believed that they had found techniques that got close to meeting these requirements (see Chapter 3). As often proves to be the case, technology that has been developed for either military or emergency service purposes is subsequently found to have all kinds of 'spin-offs' for consumer applications elsewhere. Hence, given that the development of these potentially costly positioning techniques was already completed for regulatory reasons, LBSs now had the stable platform that they needed for the development of local consumer services to take off at reasonable prices.

4.2.2 Emergence of the Third Generation Partnership Project (3GPP)

By the second half of the 1990s, North American participation in the ETSI SMG committee had become a regular, and accepted, phenomenon. Officially, these delegates initially only had observer status in ETSI meetings, although the ETSI rules of procedure were being broadened at the time to enable people outside Europe to participate at technical meetings as associate members. It soon became clear that in selecting the radio access technique to be used in UMTS, delegates not only from the USA, but also from Japan and Korea, wished to participate in the ETSI selection process for UMTS radio systems. It was during early 1998, in Paris at ETSI SMG#24bis, that the famous decision was taken to use W-CDMA as the radio system for UMTS.

Associate membership status of ETSI effectively left the Americans, the Japanese and the Koreans feeling like second-class citizens in ETSI SMG, which was not viable for any longer term future for successful collaboration on UMTS. For other reasons, beyond the scope of this chapter, the ETSI General Assembly (GA) at the time did not want to totally broaden out its full membership category to non-Europeans. Hence, a different arrangement had to be found for Europeans to work as equals alongside Americans, Japanese and Koreans. To cut a long story short, after much argument, the negotiations led to the creation of 3GPP at the end of 1998 with a new election process for the 3GPP leadership that was completed by the middle of 1999.

The Chinese authorities, through their CWTS organisation, also soon indicated that they wanted to join 3GPP. It is important to remember that China is still the largest GSM market in the world. Despite the relatively low penetration of GSM in China, there are so many Chinese that they still have more mobiles than any other nation, offering plenty of scope for the growth of their mobile market. New users in China are increasing at a rate of five million per month. Hence having the CWTS join 3GPP was seen as a significant strengthening of the fledgling 3GPP organisation in late 1999. Therefore the six organisational partners (OPs)[2] in 3GPP responsible for the development of UMTS technical specifications (TSs) are:

- ETSI (for Europe);
- ANSI T1 for the USA;
- both ARIB and TTC for Japan;
- TTA for Korea;
- CWTS for China.

4.2.3 Standards versus 'Technical Specifications'

In order to keep the 3GPP UMTS standards/specifications adopted to be the same globally, it soon became clear that any public enquiry (PE) phases in the UK and elsewhere in Europe for new 3G standards/specifications would be unacceptable for UMTS. Any PE would inevitably introduce changes to the agreed 3GPP documents which would then stand little chance of all being universally agreed in 3GPP itself. To get round the legal and regulatory needs to hold PEs through bodies such as our own DTI/BSI, it was essential that the new agreed UMTS documents would legally have to be known as technical specifications (TSs) and not as standards.

Initially, in 1998/99, it was only the GSM North American community who were represented in 3GPP, but it soon became clear that the digital advanced mobile phone system (AMPS) community, who had adopted EDGE (enhanced data rates

[2] Elsewhere readers will also see these 'OPs' described as 'standards development organisations (SDOs)'.

for GSM evolution) technology, also wanted to come into 3GPP. The digital AMPS community by this stage were organised through the Universal Wireless Communication Consortium (UWCC) and their new technological solution is sometimes known as 'Generation 2.5' and in the USA as 'UWC 136'. After negotiations held during 2000, it was agreed that all of the remaining residue of GSM work would be taken out of the ETSI SMG committee and transferred into 3GPP, so as to enable the UWCC community to participate in the specifications development process more fully. Consequently, in the middle of 2000, the time had come for ETSI SMG to be closed down after all its remaining work had been transferred to 3GPP and its remaining GSM standards had all been converted to TSs.

4.2.4 Positioning the Mobile Internet

The fixed network Internet started to take off in a big way in the mid-1990s. Many believed that users needed to get experience and confidence in using features like financial services on the fixed network Internet before the market would be ready for the mobile Internet of 3G to take off. However, given the trend of making smaller and smaller mobile handsets (including for UMTS and 3G), questions have had to be asked as to whether the mobile Internet could really be expected to take off with these small devices. In comparison, the fixed network Internet has a lot of advantages in terms of speed and the much larger quantities of information that can be displayed at one time on a computer screen, laptop, TV screen, etc. For the mobile Internet to take off successfully it cannot simply be a mirror image of its fixed network counterpart.

It may be possible to lessen these problems using voice recognition techniques, voice browsing, and audio feedback, etc — which are currently being investigated in BT's project Gabrielle [1]. Nevertheless the advantages of bigger screens remain. This has led to the school of thought that the mobile Internet must be thought of as something totally different from, but complementary to, the fixed network Internet — and should not be thought of as a competitor/mirror image of the other. A key differentiator, it is argued, would be the provision of LBSs on the mobile Internet from UMTS terminals. An LBS should be attractive to the business travelling community, for example, when they first arrive at new destinations. Hence, the provision of LBSs is seen as an important part of the technical specification for UMTS.

Figure 4.1 summarises the way the 3GPP organisation has developed. 3GPP has set up a Future Evolution Ad Hoc Group to report to the Systems Aspects (SA) Plenary Group. Apart from this development, the 3GPP organisational structure is now stable, whereas other newer mobility groups such as the Open Mobile Alliance (OMA) — see section 4.2.6 — still have a fast-evolving organisational structure. It is thus not possible to provide an equivalent to Fig 4.1 for the OMA. The five Plenary Groups (otherwise known as Technical Specification Groups (TSGs)) in 3GPP are:

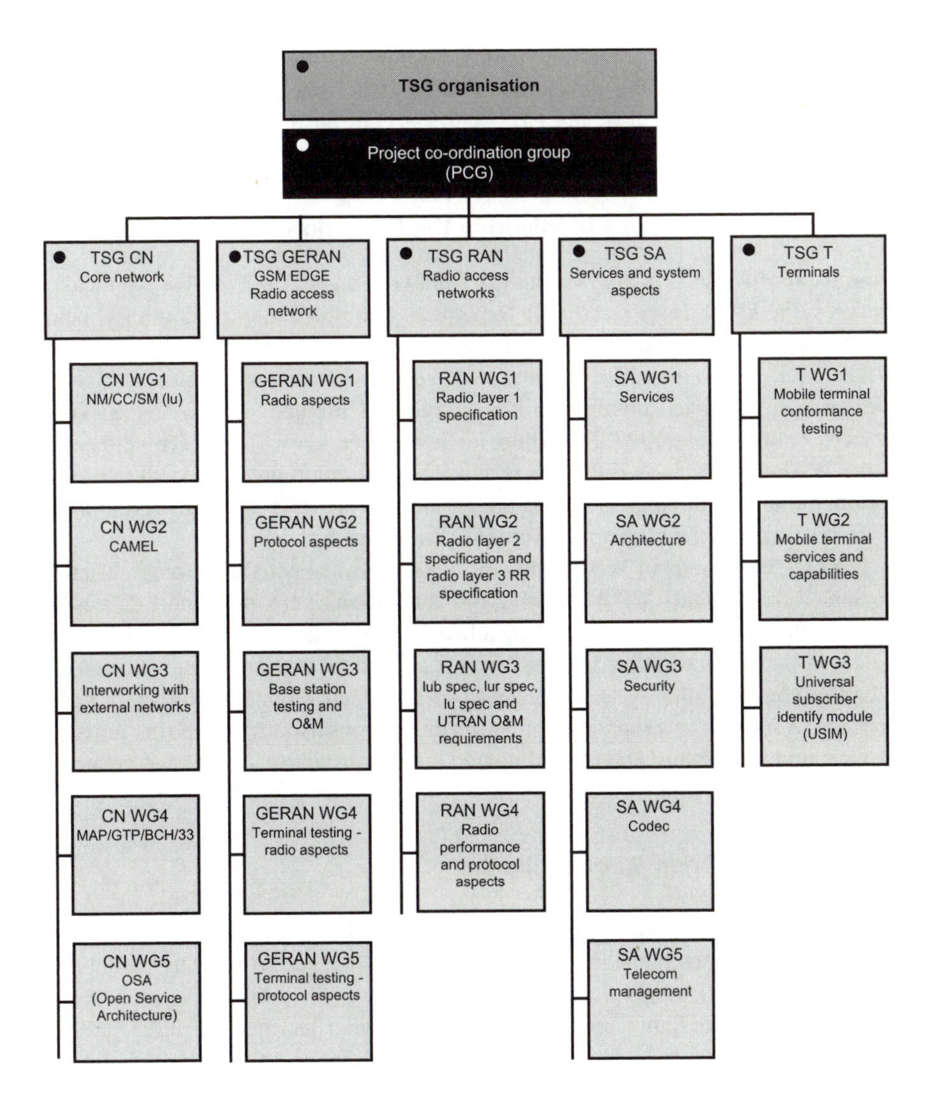

Fig 4.1 3GPP organisation.

- the Systems Aspects (SA) Plenary — includes overall co-ordination aspects for 3GPP;
- the Core Networks (CN) Plenary;
- the Terminals (T) Plenary;
- the Radio Access Networks (RAN) Plenary;
- the GSM and EDGE Radio Access Networks (GERAN) Plenary.

Arguably the third generation partnership in 3GPP is still the single most important initiative in the development of UMTS specifications. 3GPP's work includes the development of the LBS, but there are also a significant number of other international standards groups involved in the LBS development process.

4.2.5 Other Groups Involved in the Provision of LBS Standards

This chapter focuses primarily on the work of 3GPP and of the Wireless Application Protocol (WAP) Forum. It should be noted that the latter is currently being subsumed in a new initiative known as the Open Mobile Alliance (section 4.2.6).

The ETSI SMG committee and the WAP Forum started working closely together over the provision of mobile terminal end-to-end protocols for enhanced GSM services as far back as 1997. BTCellnet (as it was then known) has been very active in the WAP Forum. In 3GPP there is active co-operation not only with the WAP Forum, but also with other standards bodies such as the IETF over the development of the all-IP-based network option for UMTS releases 5 and 6.

The earlier versions of WAP protocols did not always get a good press. There is agreement that the later WAP version 2.0 specifications were much more successful [2] in pulling all the ends together into a robust and viable protocol system. However, bad press whether deserved or not can lead to 'branding' damage, and this is partly what led initially to the creation of the Open Mobile Architecture group from the WAP Forum. There were then further proposals to enhance this initiative by attracting many more players and related interest groups to sign up and create the new Open Mobile Alliance.

4.2.6 The Open Mobile Alliance

The OMA [3] had its first meeting in June 2002 and was attended by no less than 200 key organisations in the mobile industry. There had been a growing feeling that the mobility applications industry was becoming far too fragmented to be successful, and that future co-operation between what had been different interest groups was the best way forward to ensure good future market growth for 3G mobile. There is seen to be little scope for small-scale individual initiatives that are not co-ordinated within the 3G mobile industry as a whole. Competition within the industry is clearly required for success, but that is best achieved through other differentiators once a common framework of standards for areas like LBSs and mobile commerce has been agreed. The charter for the Open Mobile Alliance is to:

- deliver responsive and high-quality open standards and specifications based upon market and customer requirements;
- establish centres of excellence for best practices and conduct interoperability testing (IOT), including multi-standard interoperability to ensure seamless user experience;

- create and promote a common industry view on an architectural framework;
- be the catalyst for the consolidation of standards forums and work in conjunction with other existing standards organisations and groups such as IETF, 3GPP, 3GPP2 and the W3C.

The OMA will collect market requirements and define specifications designed to remove barriers to interoperability. It will also accelerate the development and adoption of a variety of new and enhanced mobile information, communication and entertainment services and applications. The definition of these common specifications, and the testing of interoperability, will promote competition through innovation and differentiation, while safeguarding the interoperability of mobile services across markets, terminals and operators, throughout the entire value chain.

4.2.7 Other Forums Related to the OMA

The work previously handled within the Open Mobile Architecture initiative and the WAP Forum will be continued within the new Open Mobile Alliance. The WAP Forum members at their annual general meeting in June 2002, ratified changes in their articles to fully support the creation of the OMA. Additionally, the Location Interoperability Forum (LIF), the MMS Interoperability group (MMS-IOP), SyncML Initiative Ltd and the Wireless Village initiative, have announced their intent to work with the OMA through signing memoranda of understanding (MoUs). For the development of LBSs, the inclusion of the LIF among the organisations wanting to sign an MoU with the OMA is of particular significance because this incorporates the strong vendor support from Ericsson and Nokia that LIF have, and should bring two diverging standards together.

Other industry forums focusing on mobile service specifications are also considering their future position over OMA. Of particular note (where BT already has membership) is Radicchio who would like to co-operate with the OMA, possibly also through an MoU. Radicchio is an open membership interest group [2] consisting of a wide range of companies in the mobile and financial industries. Radicchio studies solutions (both technically and business oriented) to help nurture the commercial success of mobile commerce services. The strategic importance of the OMA in the development of a wide range of 3G applications services (including LBSs) over the next few years looks very high.

4.3 LBS Work in 3GPP on UMTS Standards

4.3.1 Organisation of Work in 3GPP

3GPP is organised into five main Plenary Groups known as the 'Technical Specification Groups' (see section 4.2.4). Four of these TSGs focus on the development of UMTS specifications while the fifth (TSG GERAN) focuses on the continued development of enhanced GSM specifications, particularly for the American digital AMPS community when implementing EDGE technology for 'generation 2.5' specifications. As explained earlier, LBSs are certainly important for GSM generation 2.5 in the USA market-place. However, this chapter focuses on the development of location-based services for UMTS, where the greatest usage is expected to be in the UK and other European markets.

In 3GPP, most new work items are started in the Services and Architecture Working Groups (WGs) known as SA1 and SA2. Following this initial stage (which often includes a feasibility study), the work then fans out into other parts of 3GPP including the Security and Network Management WGs, the Terminals and Smartcard WGs and onwards into the TSG on Core Networks for UMTS and the Radio Access Network TSG. To date, most of the work in 3GPP on the development of specifications for UMTS LBSs has been at the SA1 and SA2 level on 3G location-based services and architecture — although other groups, such as the Security WG in 3GPP, are starting to show an interest in LBSs as well.

4.3.2 Location-Based Services

The range and variety of location services is considerable. The location information can be used by the mobile operator or service provider to determine billing information, such as whether the user is in a home zone or roaming overseas. In addition, emergency services and lawful interception services will want information on a user's location at a specific point of time. Typical services will include the following:

- fleet and asset management — these services typically enable a delivery company to schedule their work and predict delivery times, and it is also possible to locate animals and children or company assets, etc;
- navigation services can provide directional information in a variety of forms such as maps, verbal instructions or text messages;
- city sightseeing has been proposed as a service that is specific to the user location;
- other information, such as nearest bank, airport, bus terminal, restaurant, restroom facility, can be requested, possibly with some element of preference such as Indian or Chinese restaurant;
- more specific information may be available through a localised mobile Yellow Pages;
- the broadcasting of information to users within a geographic area typically for general advertising or to a specific group can be time specific, such as '30% off for today only'.

4.3.3 Location Methods

The methods used to determine geographic location are defined by the 3GPP RAN group. The systems and protocols used are defined in TS 25.305.

Although the techniques that use the radio signal to determine position could be based on either signal strength or round trip time (RTT) it is the latter that is currently used. Knowledge of the geographic location of the electrical centre of the transmitter masts is required when using radio techniques to determine position.

Essentially there are three current methods used for determining user location — cellID, observed time difference of arrival (OTDOA) and the global positioning system (GPS). A new satellite navigation system, Galileo, is to be introduced by the EU in 2008, representing a fourth option for determining geographic position.

4.3.3.1 The CellID System

The cellID system positions the user equipment (UE) within the coverage of the serving cell. This information is of very limited value for a large cell, but viable information may be provided for a picocell with a range of about 100 m. A further consideration is that the serving cell may not be the closest cell. It may be in the handover margin or the UE may be receiving a weaker signal from a nearer cell, due to geographic conditions or differing base-station power.

The accuracy of this technique can be improved by using enhanced cellID measurements. In this case the serving cell measures the RTT via the UE and calculates the distance from it. Again the information from a large cell is of limited value, but from a small cell the accuracy can be in the order of tens of metres.

4.3.3.2 Observed Time Difference of Arrival (OTDOA)

The OTDOA system uses triangulation from at least three base-stations to determine geographic position. The distance from each station is again calculated from the RTT via the UE. The accuracy achieved by this system is in the order of 100 m.

A fundamental characteristic of the CDMA radio system is that, if a UE is close to the serving base-station, it can have problems hearing the other base-stations on the same frequency. Since measurements from at least three base-stations are required, a solution to this 'hearability' has to be introduced. The technique used is the introduction of idle periods. During these idle periods the transmission of the serving base-station ceases for a short period of time to enable the UE to 'hear' the other base-stations.

4.3.3.3 Global Positioning System

The American military global positioning system uses satellite signals to determine a user's location. It is the most accurate technology available, typically providing resolutions of less than 10 m. The downside of using a GPS-enabled UE is the cost, increased battery consumption and data-acquisition time. The battery consumption and data-acquisition time can both be reduced by the use of network-assisted GPS. In this mode much of the information required by the GPS unit in the UE is provided by the network and so start-up and acquisition times are reduced giving a corresponding reduction in battery consumption. The network assistance will also increase the sensitivity of the GPS unit and enable it to operate in environments with a poor signal-to-noise ratio.

4.3.3.4 The Galileo System

Galileo is the new satellite navigation system sponsored by the EU and is being undertaken by the European Space Agency. It is scheduled to come on-line during 2008. It has some significant advantages over GPS in that:

- it has been designed as a non-military system and so can provide a high level of continuity of service, uninterrupted by military requirements;
- the constellation of satellites is designed to provide a potentially higher level of accuracy than GPS and to extend coverage to areas at extreme latitude;
- it sends an integrity signal informing users immediately of any errors that occur;
- it will complement GPS since a receiver will typically work on both GPS and Galileo signals — this will result in enhanced precision and a higher confidence of service continuity.

3GPP is including the Galileo system in its standards as one of the technologies that will be used to determine the geographic position of a user to enable location services in UMTS release 6. Galileo is also a source of high-precision timing information. It is believed that this timing information can be used to enhance the synchronisation of the UMTS base-stations and hence increase the capacity of the system.

4.3.3.5 Other Points

Some of the systems can also provide information on a user's altitude. This service can typically be used to identify on which floor of a building the user is located.

Additional calculations on the location data can also provide information on the velocity and heading of the user. The heading information is used in navigation systems, both to guide the user and to align maps to the heading.

There would appear to be potential for the velocity/heading information to be used by emergency services to determine which lane a vehicle was using in the event of an accident (see Chapter 3 for more details on call location for emergency services). There are reported to be systems in development that will send out an emergency message, including location information, via a cellular radio when an airbag is activated. If an emergency vehicle approaches an accident in the wrong lane of a motorway or dual carriageway, then it will have to change lanes at the nearest available exit, which may be some significant distance from the accident. To determine the position of an accident solely by horizontal positioning would require a high degree of accuracy, but a more moderate horizontal accuracy requirement coupled with velocity or heading data would appear to be a better option. It should be noted that emergency services do not currently require velocity data.

4.3.4 Service Standards for Location-Based Services

These requirements are being generated in the SA1 Services Group in 3GPP through specification TS 22.071. The current topics associated with LBSs include privacy and accuracy information.

- Privacy

  The requirement for personal privacy requires that the disclosure of a user's location to a third party should be at the discretion of the user. There are, however, four categories of location request that do not require a privacy check:

  — lawful interception;

  — emergency calls;

  — by the serving network for anonymous tracking for statistical purposes;

  — by the home network as requested by the home network operator for its own internal purposes.

  The introduction of an anonymous privacy class is also being proposed. This will enable a user to gain access to local information without disclosing their identity. The concept of an organisation anonymously requesting location information on users will be considered in the future, but may meet opposition.

- Accuracy information

  The concept of specifying accuracy requirements when sending a location request, and that of providing information on the accuracy achieved, is also being investigated.

4.3.5 Architecture for Location-Based Services

3GPP has been working on the LBS specifications for several years now and the key one for architecture is TS 23.271 which contains a functional description for LBS architecture.

Figure 4.2 shows the general arrangement of the location service feature in GSM and UMTS. This illustrates, generally, the relationship of LBS clients and servers in the core network with the GSM (i.e. GERAN) and UMTS (i.e. UTRAN) radio access networks. The LBS entities, within the access network, communicate with the core network (CN) across the A, Gb and Iu interfaces. Communications between the access network LBS entities make use of the messaging and signalling capabilities of the access network.

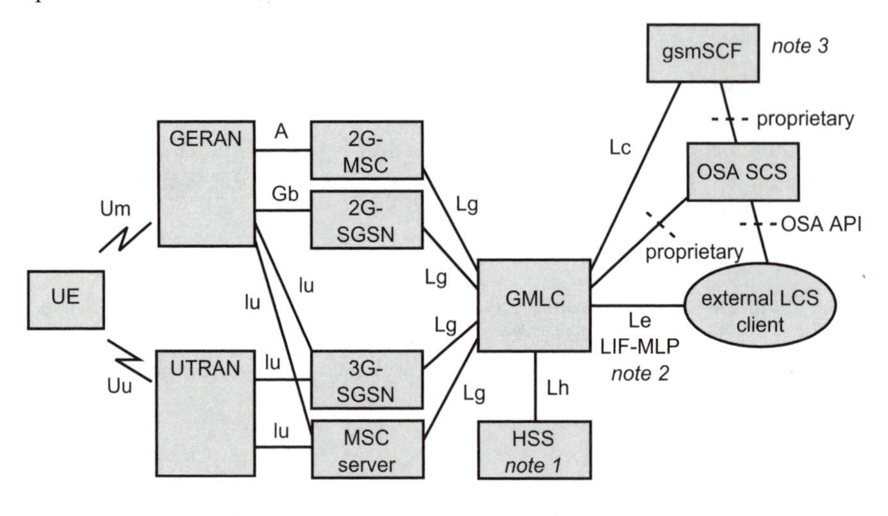

gsmSCF - GSM switching control function,
GMLC - gateway mobile location centre,
HSS - home subscriber server,
MSC - mobile switching centre,
OSA - open systems architecture,
SCS - service capability server,
SGSN - serving GPRS support node.

Note 1: HSS includes both 2G-HLR and 3G-HLR functionality. LBS is included in the overall network architecture in TS 23.002.

Note 2: LIF-MLP may be used on the Le interface.

Note 3: As one alternative, the LBS client may get location information directly from GMLC, which may contain OSA Mobility SCS with support for the OSA user location interfaces (TS 23.127 and TS 29.198).

Fig 4.2 General arrangement of location-based services.

As part of their service or operation, the LBS clients may request the location information of user equipment. There may be more than one LBS client. These may be associated with the GSM/UMTS networks or the access networks operated as part of a UE application or accessed by the UE through its access to an application (e.g. through the Internet).

The clients make their requests to an LBS server. There may be more than one LBS server. The client must be authenticated and the resources of the network must be co-ordinated, including the UE and the calculation functions, to estimate the location of the UE with the result being returned to the client. As part of this process, information from other systems (other access networks) can be used. As part of the location information returned to the client, an estimate of the accuracy of the estimate and the time-of-day the measurement was made may be provided.

4.4 Work in the WAP Forum on Location-Based Services

As explained earlier, the WAP Forum was re-organised into the OMA in June 2002 with the basic structure remaining intact. The LBS specifications available today were mainly written during the time of the WAP Forum before the release of WAP version 2.0. They were frozen in the 12 September 2001 specification version and are planned as feature-driven releases, to provide support for location-smart services.

4.4.1 History

The WAP Location Drafting Committee (LocDC) started work on location specifications in January 2000. At that time, there were groups defining location application programming interfaces (APIs) in the context of the intelligent network (IN) (e.g. Parlay), but no other groups were addressing location for Internet services. There was considerable interest in doing this work in the WAP Forum but progress was slow.

The Location Interoperability Forum established contact with the WAP Forum in February 2001. The WAP LocDC then received a presentation and a draft of the LIF mobile location protocol (MLP) API as a WAP member's contribution. The LIF MLP 1.0 specification was used as a basis for the current data structure and services that the WAP location interface has today. Unfortunately, the dialogue between LIF and WAP could only be one-sided since no IPR licensing agreement was in place and it was considered by the WAP Forum to be legally unsafe (i.e. information could come from LIF but no comments could be returned). Thus, without formal relationships and liaison in place, the protocols evolved independently into separate branches.

Future direction and operator alignment with LIF are now being discussed with potential concerns being the existence of two divergent standards (agreed to be undesirable) and the potential ability to bypass the operator's location infrastructure. While the initial specifications are complete, the current situation is confusing and needs resolution as there is an urgent need for location standardisation.

The formation of the OMA has helped the move towards convergence with LIF via an MoU and will lead to a solution to the divergent standards problem in the short/medium term, with harmonisation work likely to begin in the near future.

4.4.2 Content and Current Status

The WAP Forum location specifications consist of:

- location framework;
- location protocols;
- location XML document format.

XML-based technology is used to represent and transfer location information, due to the speed with which it can be developed and its capability for richer data, enhancing the first 3G location-based services.

The content is relatively simple as issues such as privacy, context, personalisation and positioning technologies were considered out of scope and separate technologies are being created to deal with presence.

The WAP Forum has produced a requirements document and a client/server location framework from which have evolved protocols including the XML document format specifications. At present it defines three services (described below) — query, attachment and external interface (EFI) functionality to communicate, request and reply to a location query using XML document exchange on hypertext transfer protocol (HTTP), wireless session protocol (WSP), and WAP PUSH.

- Location query service

  The location query service allows an application to query the WAP location query functionality for the location of a WAP client. An example application that could use the location query service is a tracking application (e.g. for fleet management).

- Location attachment service

  The location attachment service attaches location information to a WAP client request. An example application that could use the location attachment service is a 'find the nearest restaurant' application.

- EFI location service

  The EFI location service allows a WAP client to request the location via an external functional interface in the terminal. An example application that could use the EFI location service is a WAP client-based navigation application.

Queries can be immediate or deferred, for example in tracking applications, where update requirements can be instant or periodic. Location information consists of elements for location, shape, quality of position and triggers for the queries.

4.4.3 The Differences with LIF and Future Convergence

The LIF was formed by Motorola, Nokia and Ericsson and created an MLP defining three levels of service category — basic, enhanced and extended [4]. It has a modular XML API, for interfacing between the location server and the application and querying mobile device position independently of the underlying network technology. It runs on HTTP/1.1 and is compatible with the current WAP location framework.

The WAP location specification covers HTTP, WSP and PUSH — all as possible transport mechanisms. The LIF specification covers HTTP transport, though it indicates that WSP and simple object access protocol (SOAP) mappings will be provided in the future.

Both specifications include a similar basic query mechanism, although WAP location DC has reformulated the deferred query service and most of the draft technical documents. The attachment service is not currently covered by the LIF specification.

4.4.4 Future Impacts

Despite the standards confusion, operators have an urgent requirement to roll out LBSs, and at least one UK mobile operator is using the Redknee Synaxis Enhanced Location Server (ELS) complete solution to develop and expand their range of location-based services. This provides mobile location management functionality with Web portal and voice portal capabilities for 2G, 2.5G and 3G networks. The system queries the home location register for cellID location information and uses a SOAP XML interface to standard APIs allowing control over information privacy, content and profiling, with support for roaming and presence management.

These location standards are likely to have an impact generally on the IPR of many project/product areas within BT such as:

- uLoc8™ (personalised and location-sensitive application);
- Rocking Frog (providing timely, context-sensitive, personalised data to users with unique searching filtered on relevance to their interest profile, location, device used, time and motivations);
- Wireless Propositions (specific location-based service components for providing consultancy);
- Erica (mobile applications enabler for mobile operators, providing an API between the application and the underlying network);

- Foresight (coupling positioning technologies with ubiquitous devices and virtual models of the real world to enable a much richer interaction between people and systems);
- Event Management Systems (for information gathering, management and transmission onboard vehicles during races/exhibitions, e.g. yachts, cars);
- Personalisation Venture (personal profile structure and hosting software);
- Registered Service (for a personalised profile — non-anonymous).

4.5 Summary

Any future mobile applications, which require the transfer of location or indeed personalisation information in a standardised way, are likely to use the location protocols outlined above. It remains to be seen whether convergence and harmonisation between the standards bodies will refine the protocols to a more sophisticated form with the urgency which is required, but the indications are that this year will see the necessary work being done.

With the size and scope of this potential market, the key question remains as to whether any serious players can afford not to be involved in the development and maturation of this next generation of LBS standards.

References

1 Ringland, S. P. A. and Scahill, F. J.: '*Multimodality — the future of the wireless user interface*', BT Technol J, **21**(3), pp 181—191 (July 2003).

2 Radicchio: '*Wireless eBusiness Security*', Barcelona Conference (January 2002).

3 Open Mobile Alliance — http://www.openmobilealliance.org/

4 Baxter, R.: '*Mobile location protocols*', Internal BT report (March 2001).

5

PROFILING — TECHNOLOGY

R Newbould and R Collingridge

5.1 Introduction

Stepping into cyberspace has become a frequent activity for many of us for work and leisure, and we are coming to depend on it increasingly as the electronic environments become more useful and immersive and we gain access to higher bandwidths. For the purposes of this chapter, cyberspace can be defined as a realm in which communication and interaction between individuals and computers are facilitated by data exchanged over computer networks [1]. The most notable examples of cyberspace are of course the Internet, intranets and extranets.

In order for the interactions in cyberspace to go beyond the simplistic, cyber-citizens require computers and individuals to have access to personal information, e.g. to personalise a news site so that it contains more relevant stories on the homepage. Some information about end users must be tied to real-world properties, such as credit card details and addresses for despatching goods, or, for more sophisticated applications, the customer's current location to notify the traveller of the whereabouts of particular services. Other interactions may be divorced from reality, such as the choice of an avatar and personality for participating in a virtual world. A demand is anticipated for ever more detailed information about users such as calendar information to help people organise holidays with a group of friends.

All properties used to personalise interactions with computers and individuals constitute a user profile, and techologies for hosting user profile data are the subject of this chapter. There are a number of factors shaping the architecture of such technologies, including technology enablers, the key customer and business benefits, the need for identity management systems and business imperatives, all of which will be discussed in section 5.2. A number of initiatives will then be considered in the light of these factors, to analyse the benefits offered and which are likely to be the more successful strategies. Section 5.4 focuses on how some of the key issues were addressed in developing a profile hosting server (PHS), delving into the design to yield insights into the technical composition of such systems.

5.2 Drivers

5.2.1 Technology Enablers

The birth of cyberspace, as we know it today, came with the Internet, which became popular with the masses in the 1990s following the evolution of the World Wide Web (based on hypertext transfer protocol and hypertext markup language). More recently, XML (extensible markup language) has aided the definition of inter-machine communications, facilitating the development of advanced capabilities. In the last year, Web Services technology has burst on to the scene, building on XML, and is a key enabler for more complex distributed systems including the sharing of user profile information.

Web Services [2] promise to revolutionise computer-to-computer interaction by simplifying integration, analogous to how the World Wide Web facilitates human/computer interaction.

For the purposes of this chapter, a Web Service interface can be defined as one employing, as a minimum, Web Services description language (WSDL) and simple object access protocol (SOAP). The former protocol describes the interface and the latter invokes it.

The Web Services standards are important for profile hosting as they enable parts of a profile to be accessible by multiple consumers, such as GSM cell location-to-location-tracking Web sites. Higher layers are being added to the Web Services standards stack to specify security and business process automation which will enable powerful trusted applications using peoples' profiles, a glimpse of which is revealed in the scenario following.

5.2.2 Profiling Benefits

The introduction illustrated a number of cases where service to users is enhanced through access to profile information, which is accrued today by on-line sites, explicitly through on-line forms and implicitly through behaviour analysis.

Figures 5.1 and 5.2 depict a travelling salesperson scenario illustrating the power of being able to hook Web Service-enabled capabilities together, perhaps run by different companies, utilising communications service provider resources and centralised, managed user profiles.

The main benefits illustrated for the end user are 'taking the hassle out' of tasks and even making new capabilities possible due to 'interconnectedness', such as locating and contacting people within a short timeframe. For businesses, there are benefits through more effective and therefore profitable targetted selling, and more satisfied and therefore loyal customers. However, the rights of individuals must be protected from abuse of their profile data.

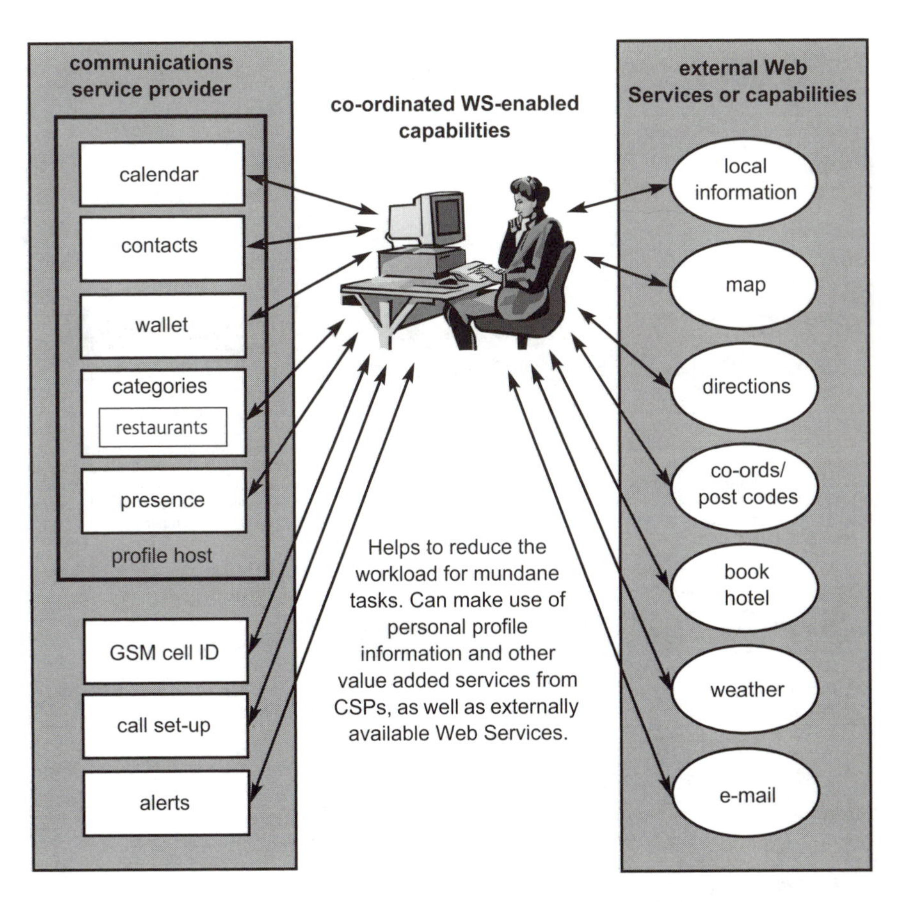

Fig 5.1 Travelling salesman scenario — Web Services capabilities.

5.2.3 Network Identity Mangagement

The things that are essential or unique about people are what identify them. For example, we might use hair colour, whether they are over eighteen, an identification number or, more commonly, a name. Network identity is the total set of characteristics about a person held on a computer network, and since any part of a profile could be used to identify a person for some purpose or other, the profile in conjunction with a profile ID, constitute a person's network identity.

In any real-world interaction, there are always identifying traits that come bundled with an individual. For example, simply walking into a shop reveals height, indications of income, and so on, whereas in cyberspace it is possible to unbundle characteristics in such a way that individuals need reveal only those they wish to [1]. To enable users to protect their own privacy from parties who might use their

1 Travelling salesman wishes to organise evening. Firstly, the service works out where he will be from his calendar or form entry. It generates a list of hotels using a 'local info' service. He can click 'Phone' to automatically set up an IP-PSTN voice call, or, if the hotel is WS-enabled, book it directly, utilising profile wallet data.

2 Map is returned automatically, showing location of hotel, using a mapping service.

3 He is now presented with a list of possible restaurants, based on his preferred food profile, located within walking distance of the hotel (determined by using the postcode to co-ordinate mapping Web Service). He may book a table using the same means as for the hotel.

4 Web Services directions for walking from the hotel to restaurant, and a weather forecast, indicating whether to carry an umbrella.

5 Finally, the service determines which contacts may be in the vicinity by searching the profile contacts for addresses, the contacts' calendars and even their current location (e.g. from cellID). He selects those to invite to the meal and a message is auto-generated. The message is sent using an alerting mechanism if available. Alerting mechanisms use 'presence' information such as whether the person is currently on-line or off-line, busy or interruptible, typing or possibly away from the computer, etc. Rules can determine the best means for contacting individuals, such as via SMS, an Instant Message, etc.

Fig 5.2 Travelling salesman scenario — Web Services in use.

identity information in a harmful or intrusive way, network identity architectures will, no doubt, evolve to enable users to give away the minimum of information (architecture and privacy pressure).

At the same time, identity information has become a commodity that service providers wish to obtain in order to market their goods more effectively. They prefer the maximum bundling of characteristics, and try achieving this by providing value to the end user for sharing the information, including financial incentives. Figure 5.3 illustrates this tension. An appropriate balance should be found naturally, though if commercial influences push this too far to the left for the good of the public, governments may have to step in with new legislation.

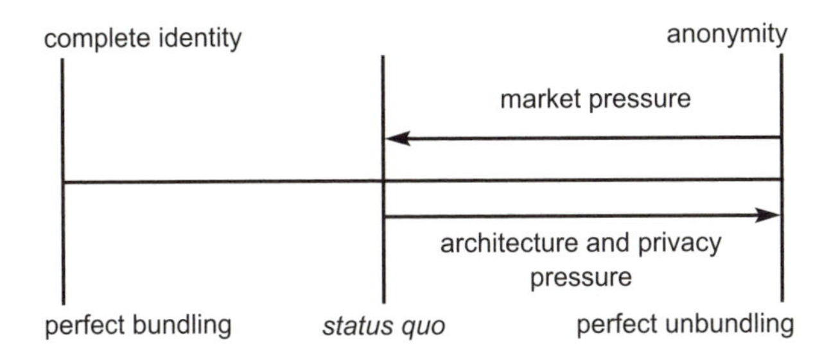

Fig 5.3 Bundling versus unbundling of personal characteristics (adapted from Abelson and Lessig [1]).

Anticipating a rising awareness in users for the need to manage identity information held in computer systems, Alan Westin, a leading thinker in the privacy arena, made the following prediction: '... most individuals will consider and assert privacy choices as a common daily practice.' He also observes that '... informed consumer choice is the core of the new privacy paradigm' [3].

Clearly, users will need some method to control and track what information is released to service providers in cyberspace. Additionally, all parties in cyberspace potentially require a system to verify identity assertions, be they for business-to-consumer, business-to-business or user-to-user interactions. This spells the need for a network identity infrastructure, which is able to tie people to the real world but offer the capability of unbundling. Such a system has a strong synergy with single sign-on requirements and so it is not surprising that efforts in this area tend to associate the two (see section 5.3). Governments should also have an interest in this area, as there is arguably a need to be able to trace actions down to individuals in order to protect society from an explosion of crime taking place on the Internet.

The following section considers some commercial issues to do with the provision of such a network identity management system.

5.2.4 Commercial Issues

A network identity infrastructure ought to be open in order to derive the most benefit, since there is rising marginal value for each user and service provider on the system as it will tend to increase the total number of potential customers and the quality and size of personal profiles. Herein lies a difficulty — no company would want to foot the bill for developing a truly open standard that would benefit the rest of society, as companies would rather capture the whole value themselves. In fact, there is a tendency to keep identity management systems closed in order to do this — hence most initiatives have been proprietary until recently.

In addition, companies implementing network identity systems obtain business advantage from increasing the amount of profile information they can capture; hence end users may lose some of the unbundling power that would help protect their privacy. However, end users are unused to paying for such identity systems themselves. It seems likely that the highest costs associated with network identities will be that of verifying credentials [4] where necessary, such as age or address, which could be guaranteed through public key cryptography schemes.

As the following look at standards and technologies reveals, there is considerable business interest in developing network identity standards and infrastructures to benefit on-line sites while providing a degree of privacy protection to keep users happy. The commercial issues presented here have proved to be significant barriers that so far have not been entirely addressed by these initiatives.

5.3 Standards and Technologies

5.3.1 P3P

The platform for privacy preferences (P3P) project [5] enables Web sites to express their privacy practices in a standard machine-readable format that can be retrieved automatically and interpreted easily by user agents, thus removing the need for users to read site policies and enter information manually each time. P3P user agents, such as Microsoft's IE6 P3P add-in, allow users to be informed of site practices and to automate decision-making based on these. For example, users may set up a profile and sharing rules including physical contact information that they are happy for their agents to automatically share with any organisation that will not pass it on to third parties, and medical information for which they require a user prompt before sharing.

P3P goes some way towards providing users with control over profile sharing but there are a few outstanding issues with the approach. Firstly, it is debatable whether users would want to control the release of information based on privacy policies alone, especially since P3P offers no mechanism for ensuring adherence to the policies. It is perhaps more likely that users would base decisions on their perceptions of a company's trustworthiness and the value provided. Secondly, client-based hosting of profiles limits their usefulness, as there are certain services which require access to profile data when the user is off-line, such as colleagues wanting to read a current GSM cell location. Finally, P3P does not specify a full identity management infrastructure, though such a system based on P3P has been proposed [6]. The following offering from Microsoft went some way towards addressing these issues, though in a closed manner.

5.3.2 Microsoft .Net My Services

Around April 2001, Microsoft announced an identity management offering as part of their .Net Web Services strategy, which came to be launched as .Net Passport/My Services [7]. Users must subscribe to Microsoft Passport, which is a single sign-on system used to access a number of Microsoft services including Hotmail and service providers they have signed up. The intention was that service providers could access a potentially broad user profile hosted by Microsoft on their .Net My Services platform, including preferences, eWallet, calendar and alerts. My Services would put users in control of their profiles (discussed in section 5.2.3) by displaying a Microsoft .Net My Services user interface [8] to enable users to set access rights for different service providers to different parts of their personal data (see section 5.4.1 on the BT Exact profile-hosting server for a fuller explanation of how this could be achieved).

After a number of months trying to sign up organisations, Microsoft accepted it did not have a clear enough business model for .Net My Services [9] and eventually withdrew the offering [10]. Problems reported include lack of trust over the security of Passport and resistance from organisations not wishing to allow Microsoft to host commercially valuable profile data. Microsoft also suggested making customers pay for the hosting of profile data, but this is an untested strategy, as users are not accustomed to paying for such services. In response to issues faced, Microsoft announced that it would change its strategy in two important ways. Firstly, it promised that version two of Passport will use the trusted Kerberos authentication architecture. Secondly, it claimed that My Services would support federated profile hosting, an approach also adopted by the Liberty 1.0 standard [11].

5.3.3 Liberty

The Liberty Alliance project [12] is a Sun-led alliance comprising over 60 member companies, formed to deliver and support a federated network identity solution for the Internet. The version 1.0 specifications were released in July 2002. A person's on-line identity — their personal profiles, personalised on-line configurations, buying habits and history, and shopping preferences — is administered by users and securely shared with the organisations of their choosing.

Figure 5.4 is an overview of the Liberty 1.0 architecture, which utilises Web Services for communications between providers [12]. Since it is a federated identity solution, different service providers host different parts of users' profiles as appropriate, and users can potentially choose that data held by one service provider can be shared with other service providers. Identity providers perform authentication and can be chained together, with the active agreement of users, to provide single sign-on.

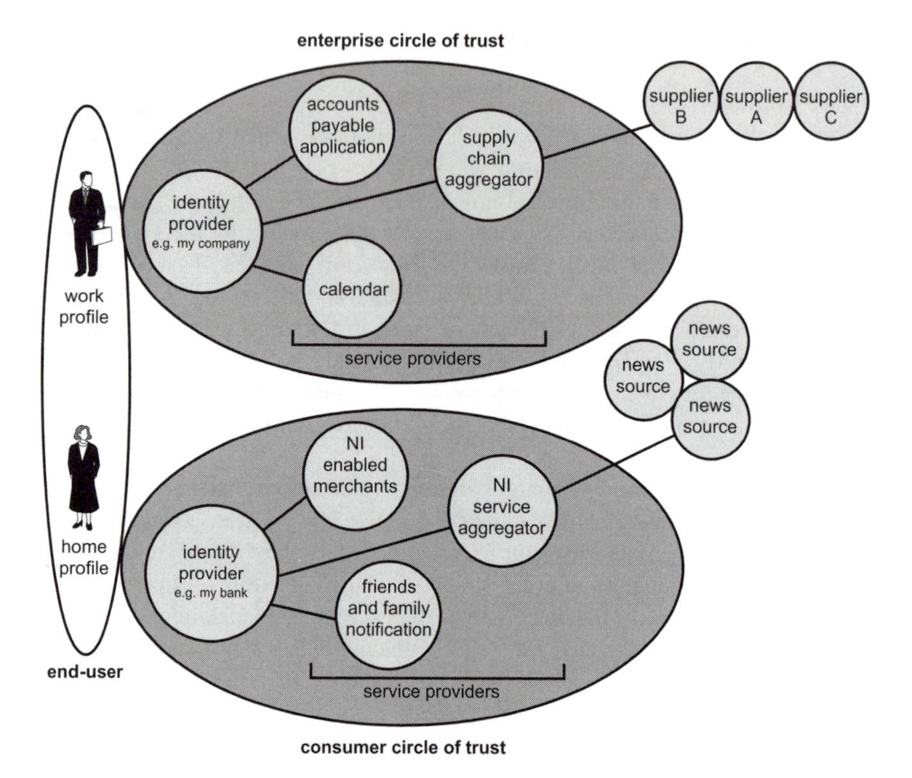

Fig 5.4 Liberty architecture — federated network identity.

The Liberty standard is an open technical specification, but users cannot link an identity provider with any service provider or share data between any service providers of their choosing. Providers can only be linked where service and identity providers have chosen to affiliate for mutual business benefit, creating circles of trust, as illustrated in Fig 5.4. The Liberty architecture does remove the need to rely on a third party to host profiles and the security risk of a single organisation hosting entire profiles. However, in its current form it falls short of the ideal network identity infrastructure, since commercial interests look likely to resist the full inter-linking necessary for global single sign-on and profile sharing on the Internet (see section 5.2.4).

5.4 The Government User Identity for Europe (GUIDE)

A number of European companies are proposing an EU 6th framework project named GUIDE to pave the way for technological and legislative measures for electronic identity to be provisioned and maintained throughout the life cycle of

European citizens, for social and commercial purposes, on the Internet. This could well prove to be the best way forwards for the EU as a whole.

A government-directed scheme opens the possibility for a network identity infrastructure that truly is a 'common good,' benefiting all parties to the maximum (see section 5.2.3). It could also take the opportunity to build mechanisms into the architecture to enable perpetrators of Internet crime to be traced, although this needs to be balanced against the need to protect the privacy of civilians.

5.4.1 Profile Hosting Server

The BT Exact profile hosting server (PHS) started life as a research project in 1998, looking at methods to allow personalisation of applications and services. The stated objectives of the project have changed little since its inception and are:

- to be an enabler to personalisation for any service or application;
- to have scope across the whole Internet, including intranets, isolated networks or *ad hoc* network segments;
- to enable personalisation on any client device or multiple simultaneous devices, at any time and from any location or network — this includes simple voice services, WAP telephones, PDAs, pocket PCs, desktop PCs, set-top boxes, games consoles, kiosks and in-car services;
- to allow users to explicitly represent themselves, their attributes, and interests in one single (but possibly distributed), secure profile, hosted by a party acceptable to the user;
- to enable knowledge management and community working;
- to allow authorised applications, services and individuals to interact with a user's profile in a clear and efficient manner.

In developing a service to meet these objectives, BT has gained some valuable intellectual property rights in the personalisation and profile hosting arena, and has built considerable skills and knowledge in the personalisation space.

The objective of the following sections is not to provide a complete overview of the implementation, but to provide an insight into the issues and technical challenges faced in meeting the above objectives.

5.4.2 Profiling Issues

5.4.2.1 Privacy

Privacy is perhaps the most important aspect of profile hosting from the user perspective. It is essential that the public perception of privacy for any profile hosting service is maintained at the highest level. In order to achieve this the following conditions must be met:

- use of information must be consensual — returning to customers to ask for permission to use data for new purposes;
- information processing should be transparent — users should be able to see what data has been collected about them;
- all use of information should be fair — simplistically, users must be happy with the way in which the data about them is being used.

These messages must then also be conveyed to the user in a clear and unambiguous privacy policy, which must be presented to the user at registration time. The above is a subset of the principles outlined in the 1998 Data Protection Act [13], which is enforced by European directive 95/46/EC [14].

5.4.2.2 Security

It is essential that a user's privacy is maintained, but equally it is important that users have and maintain trust in the systems' ability to securely hold, and manage access to, their profile contents. Not only must the platform and software be secure, but also the transport and delivery of information to other parties must be highly secure. One further essential requirement is the ability to provide a detailed audit trail, so that profile access cannot be repudiated.

5.4.2.3 Ownership

To ensure user acceptance, a profile hosting service must clearly present the terms and conditions of the service provided. Right from the outset of the project it was clear that the one fundamental principle required to engender trust and respect for the user's privacy was to ensure that ownership of a user's personal information remains with the user. The concept of a user-owned profile was perhaps the most significant driver behind the design of the PHS.

Having said this, there is also a market demand for the situation where a business may want to hold and own personal profiles within a defined business scope, e.g. restrict the sharing of personal information to partners and affiliates only. A secondary consideration was how to support this model without compromising the primary goal of user-owned profiles. This challenge was successfully addressed in the development.

5.4.2.4 Scope

It is essential that the scope of the personalisation is bounded and this is usually limited to a business domain. As an example a user's shipping address may be correct and relevant within one shopping application where they purchase books for work use, but will not be relevant in another where they may purchase CDs for personal use.

The PHS is designed to support personalisation for users across the whole Internet, yet provide mechanisms to support limited scope, e.g. an intranet environment or a business-specific environment. This partitioning of business domain data is essential in maintaining privacy and business confidence in the concept of profile hosting.

5.4.2.5 Schema

It is essential that all objects being profiled to deliver personalisation use a commonly understood schema so that relationships can be made between objects that have been profiled. This places emphasis on using the correct ontologies and an architecture that allows them to evolve, as understanding of how users and businesses apply profiles improves.

The schema must be flexible to allow it to support various media types, and extensible to allow service providers to store additional information in a structure that suits their business needs.

5.4.2.6 Purpose/Context

It is essential that the meaning and purpose of all attributes within an object's profile be clearly understood, so that the attribute is not used out of context. As an example, a user may have an interest in banks but this must be captured as a financial interest and not one about rivers.

5.4.2.7 Timeliness

It is essential that the attributes of a profile are relevant in the time domain. As an example, a user may not want to receive endless offers about red wine just because they once bought a bottle as a gift in the distant past. There must be a mechanism for the 'tidy up' of profiles and the ability to remove attributes that are no longer of relevance.

5.4.3 Design Overview

With such a complex set of requirements and operating environments to consider, the PHS had to be designed as a fairly simple, low-level capability that could be used to enable personalisation in a wide range of environments. The analogy used was that the PHS must provide a profiling service comparable to that of an SMTP mail server, in that it should be simple, portable, scalable and performant, requiring minimal hardware and capable of running on multiple platforms. This is also a good analogy in that a user's profile, like their e-mail account, can be hosted by their chosen service provider and moved if the service provider does not meet their expectations of service levels.

Simplistically, PHS is a profile store or database with a security layer, to enforce privacy and security. The profile is designed to be extensible and portable. The user or hosting service provider can back it up and it can be distributed across multiple platforms. The security layer, called the gatekeeper, is described in more detail in section 5.4.5.

One of the key design decisions made at an early stage was that the permissions or access control lists associated with a user's profile are embedded within it. This decision is critical to the profile having application in an *ad hoc* network environment or a localised network environment, where access to a centralised, network-based authentication system (e.g. Microsoft Passport) would not be possible. This ensures that the profile can be used in both a network-based or client-based environment and that synchronisation can be used in a distributed profile environment.

In order to simplify integration of the PHS into existing applications and services, a decision was made to use an emerging technology, called SOAP as the transport mechanism, and XML as the access language. This was a fortuitous decision as the Web Services initiative emerged and grew over the lifetime of the project to become the implementation technology of choice today. PHS is a true Web Service with a WSDL definition.

Figure 5.5 shows a brief outline of the profile hosting service system. The components are as follows:

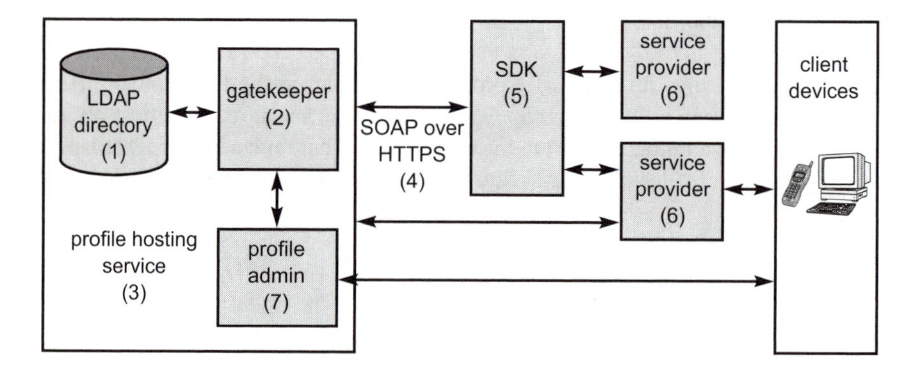

Fig 5.5 Outline of the profile hosting service.

- a lightweight directory access protocol (LDAP) directory server is used to provide the main information store for the profile hosting service, consisting of portable personal profiles (1);
- the gatekeeper turns the LDAP directory server into a profile hosting server — it provides all of the service functionality needed by the service providers to interface with the service (2);
- the profile hosting service comprises the LDAP directory and gatekeeper software — it exists on a secure network ensuring only the gatekeeper talks to the LDAP server and only the intended HTTPS (hypertext transfer protocol over secure socket layer (SSL)) requests reach the gatekeeper (3);
- the HTTPS communications system is used by the gatekeeper for communications with the outside world (4);
- the service's software development kit (SDK) provides a simple mechanism by which service providers can access the profile hosting service (5);
- service providers, as the name suggests, provide the services built upon the profile service, accessing the server by constructing and sending SOAP packets over HTTPS, or via the use of a released SDK — profile management tools also exist in the form of service providers (6);
- profile administration components are used for the creation, management and support of the subscriber's profile account (7).

5.4.4 Profile Store

Investigative work was done to see how the profile should be stored. The two key technologies considered were a relational database or an LDAP-compliant directory. The latter was chosen for the implementation as the directory schema and its access control model were a very good fit to that required for the profile. The LDAP directory also provided suitable performance and offered a high level of platform independence. LDAP directories are optimised for specific read/write ratios and testing showed that it was also a good choice in this respect for a typical customer usage profile. The modular design of the service does not preclude a migration to another storage mechanism, and investigative work into native XML databases, such as Apache Xindice [15], has been undertaken.

5.4.5 Access Control

Access control is enforced by the gatekeeper and its role is to provide access to the information stored in the personalisation system in a controlled manner. It provides this access through an HTTP interface, where the requests and responses are made up of XML documents. In providing access to the information available on the system, the gatekeeper performs various tasks such as HTTP request handling, request parsing, request fulfilment, request response generation, session tracking, caching, logging and security management.

The core profile is stored in the LDAP directory but the gatekeeper stores additional information about sessions, etc, locally. The view of the information from the client side is different to the actual storage structure used on the LDAP server. This difference in view is intended to improve ease of use for both service providers and subscribers and to present a logical model of the personalisation system, while hiding some of the detail of the implementation.

The HTTP communication is implemented following the SOAP 1.1 specification for packaging the requests and responses between a service provider and the gatekeeper. SOAP allows multiple different transports to be used including TCP, HTTPS and SMTP; moreover the gatekeeper could be made accessible over one or more of these other transports at a later date.

5.4.6 Application Interfaces

As previously mentioned, the applications interface to the service is via exposed SOAP/XML methods, which are defined and documented. In addition, a subset of the functionality is presented in the form of a set of Java classes to allow developers to write client-side applications and services that interact with profiles, and some example applications, such as tools used to analyse local data and then populate the profile based on this analysis, e.g. a user's bookmarks.

5.4.7 User Interface

For a user to accept and trust services using their profile, a clear and unambiguous interface between the PHS and the end user is required. Access to all or part of the profile is controlled directly by the profile owner. In order for a service or application to read or update the user's profile, they must first establish a trust relationship with the profile owner. A trust relationship basically involves a service requesting read or write permissions to part of the profile and the user providing credentials to the service in the form of an identifier and associated password. These credentials are stored, as part of the profile and its embedded access control list. The credentials allow a one-time access only or continued access to the parts of the profile requested. This process of negotiating access rights is called the break-out process.

The break-out process allows negotiation of access rights between the requesting service provider and the end user. It has been implemented in various forms but the most common implementation is by using HTTP redirection to allow granting of access rights to service providers.

The process is required because by default a service provider will not have permission to access all of the profile. If a request is denied, the user is informed and asked to make a decision as to what rights, if any, should be granted to the service provider. This may result in the user denying access this time, denying access now and in the future, granting access this time only or always allowing access to the requested information. Having granted or denied access to the profile, the control is then passed back to the service provider to allow the user's personalised experience to continue.

5.4.8 Profile Contents and Structure

The profile for each user is, by default, unpopulated. A basic structure is in place and it is assumed that some basic information is placed within the profile at registration. It is possible for the creation and registration process on PHS to be completed by a service or application, thus simplifying the registration process and initial profile population.

BT has also developed tools to assist in profile population by analysing other data sources. An example of this is a tool that analyses a user's bookmarks and generates an assumed set of interests from these bookmarks and the content residing behind them.

PHS supports the concepts or roles and context. Roles allow a user to split their interactions into a defined set of spaces. For each role a unique set of personal information and interests exists. This allows a user to have separate contact details and interests, for example, within a work role and a home role.

A number of contextual flags exist and these are common to all roles. This is an extensible set of attributes that reflect the user's current activity, availability, mobility, etc. There are also contextual flags with Boolean values, e.g. out-to-lunch.

Figure 5.6 shows the basic profile structure that exists on the creation of a user account on the profile server.

5.4.9 Personal Information

This node contains personal data such as name, address, date of birth, marital status, accessibility information, etc. It is held per role and follows an open standard schema as defined by the P3P W3C working group [5].

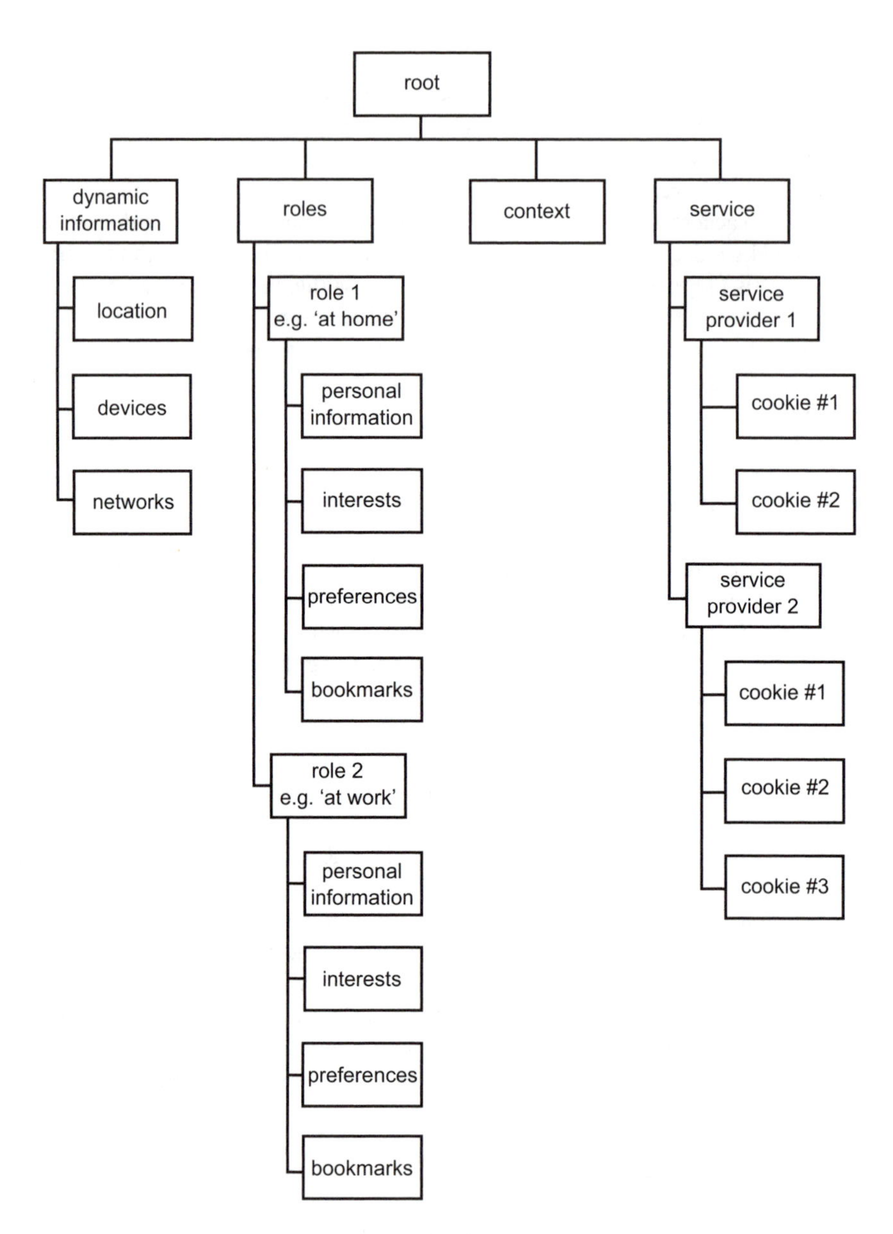

Fig 5.6 Profile schema overview.

5.4.10 Interests

For each role, a user's interests are stored as a set of weighted keywords and phrases. This format is bespoke, but is fairly common across a large number of projects within BT. There are tools available that use these to determine relevance of content to a user based on these interests and some examples are discussed later in this chapter. A user's interests evolve over time, through interaction with one or more services, and eventually become a fairly accurate reflection of the user and their interests, allowing content filtering and prioritisation to be performed to enable advanced knowledge management. PHS supports the concept of interest decay, so that an interest in a particular topic will decay over a period of months or years, unless that interest is 'refreshed' through use of services. PHS also allows storage of disinterests, so that content can be actively filtered out by applications and services, if required.

5.4.11 Preferences

Preferences are similar to interests but have a more clearly defined application and scope, e.g. favourite restaurant type. They are attributes with multiple values and weightings, allowing explicitly stated and prioritised preferences.

5.4.12 Bookmarks

This part of the user profile holds a user's bookmarks such that they can be accessed by any device or service. The profile enables bookmark management, usage tracking and linking of bookmarks to devices, e.g. 'only show me WML bookmarks when on my WAP phone'. In addition, the profile supports change notification tracking, e.g. 'highlight pages that have changed since I last visited the bookmark'.

5.4.13 Dynamic Information

The 'dynamic' information is information that changes fairly frequently, such as active device or current location. This part of the profile also allows a user's devices, network connections and preferred locations to be defined in detail and for relationships between them to be assumed, e.g. 'if I'm on my home PC, then I am at 1123 Kirton Close, Felixstowe (postcode and GPS position known) and I'm on a 64 kbit/s dial-up connection'. My 'home PC' has defined characteristics and capabilities, which can also be profiled using open standards. The PHS also stores a historical location trail and potential future locations, which may be used by services such as route planners, etc.

5.4.14 Service providers

This part of the user profile is a superset of the cookie concept. It provides a secure space for trusted parties to store information in a user's profile, to which only they have access. It is effectively a mechanism for implementing network-based cookies. The type and amount of information that can be stored is much greater though and applications have been written which use this space for session tracking across multiple clients or for portable media rights management.

5.4.15 Communications rules

This part of the user profile holds communications rules for the personalised alerting service. This is covered in section 5.5 as an example application. It allows a user to define rules for notification at the optimum time and on the optimum device.

5.5 Example Applications

5.5.1 Personalised Alerting

One of the first applications identified that would benefit significantly from personalisation and the application of a user profile was that of personalised alerting. This application has been implemented as a Web Service that trusted service providers can use to communicate with and to notify users. From a service provider point of view, the complexities of alerting are hidden. They simply invoke a Web Service with the required priority and delivery attributes set. The service uses the user profile and communications rules to determine the best way to deliver the alert to the user, that meet their chosen priorities and devices.

This simple service is linked to an SMTP mail server, an SMS gateway and an IP messaging platform, allowing alerts to arrive as e-mail, SMS messages, or pager notifications, and instant messages to a user's desktop.

5.5.2 Advanced Communities

The BT Exact Advanced Communities service is an application that combines the PHS with a range of Web Services-enabled capabilities to provide an advanced community service platform. The development focused on integrating existing technologies within BT and developing a core community engine to showcase both the PHS and the alerting engine.

Figure 5.7 shows a top-level view of the Web Services integrated to produce the community service.

The development focused on modelling communities and their users in a native XML database and in developing a community engine to allow creation and management of communities.

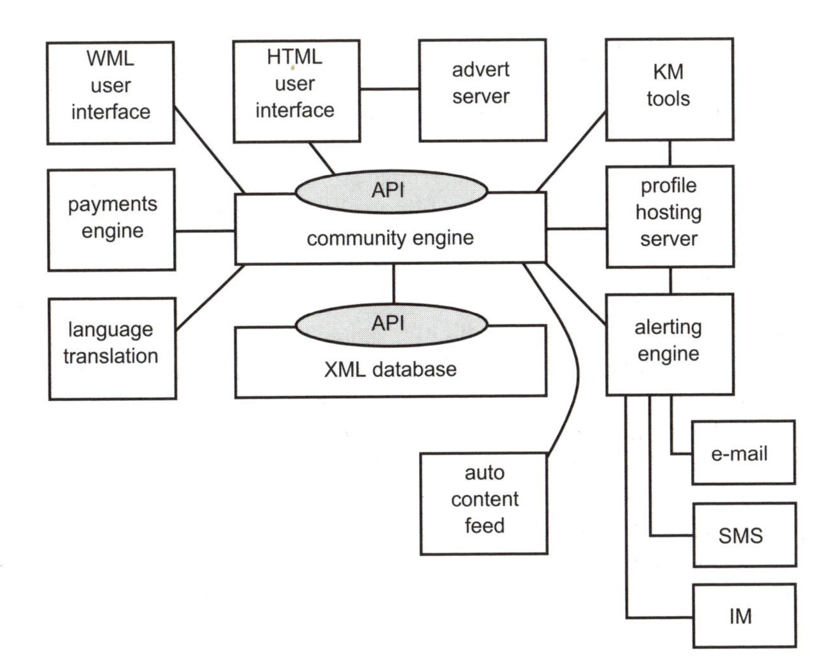

Fig 5.7 Advanced communities service.

The service is completely device independent and a user interface is required to present functionality to users. This can be a networked interface, e.g. WML or HTML, or it can be a client-based interface customised to a specific device, e.g. an XDA pocket PC. The engine supports and models the concept of communities, user roles, multimedia discussions and their moderation. The service is designed to be both cost effective and scalable from tens to millions of users.

Through advanced profiling techniques and a suite of knowledge management tools, each community, user and content item can be profiled in a common way to allow direct assessment of relevance. If shared information is deemed relevant to a community or user, they are notified via the alerting engine.

The auto-content feed allows content sources to be monitored for articles of relevance and to then be shared with the community, e.g. monitoring of USENET newsgroups of interest.

The Web Services payment engine allows users to be charged for creating communities, for subscription to communities and for pay-per-view content within communities. Since a user's preferred language is stored in their profile, the language translation service allows a community to be presented in their chosen language. The personalised advertising server is used to present advertisements targeted at both community and user interests, at the community owner's discretion.

5.6 Summary

This chapter began by exploring the key drivers behind the growing interest in, and approaches taken by, profile hosting, or identity management, solutions. Web Services is a key technology facilitating remote profile storage and has been adopted by Microsoft and Liberty Alliance profiling solutions, which represent the most significant recent developments.

The increased efficiency derived from hosting profiles centrally is of benefit to end users and businesses alike, though there is a battle between service providers and consumers over how much profile data should be bundled with basic identity information by service providers. Architectures support unbundling of profile data, allaying users' privacy fears, but businesses stand to gain from more effective marketing by obtaining as much information as possible.

Studies of recent profiling technology developments suggest users would prefer to be in control of those suppliers with whom they share data, rather than allowing an automated mechanism to decide, based on unverifiable privacy policies. This is the basis of Microsoft's .Net My Services and the Liberty Alliance's developments. The former experienced resistance to the power of a third party holding the entirety of a user's profile data, though Microsoft promises to address these issues. The latter addresses commercial issues by proposing identity federation and user opt-in for sharing of data within 'circles of trust.' Liberty's approach may not go far enough towards enabling a 'common good' identity management infrastructure that is truly open and can enable law-enforcement through tracing criminal activity. This, arguably, can only be provided with input from Government, which is where the proposed EU GUIDE initiative has the potential to add value.

BT Exact has developed the profile hosting server to meet both user and business demands for personalisation and has produced a flexible and extensible capability that can be integrated into a wide range of environments and platforms. A logical progression for profile hosting services is that of extending into the digital identity hosting space. This is a future development direction proposed for the PHS as part of a wider initiative based on open standards and Web Services technology.

Using personalisation and profile hosting, BT Exact has already produced some exciting propositions that show the potential benefits of personalisation in real world applications. Services such as Advanced Communities offer BT and its partners a real opportunity to bring compelling and addictive services to the broadband and mobile market-place and thus stimulate growth in these areas.

References

1 Abelson, H. and Lessig, L.: '*Digital identity in Cyberspace*', White paper submitted for 6.805/Law of Cyberspace: Social Protocols (Dec 1998).

2 W3C Web Services homepage — http://www.w3.org/2002/ws/

3 Westin, A. F.: '*Consumer privacy in 2010: ten predictions*', Lecture at the '*Balancing Personalisation and Privacy 2001*' Conference, Privacy and American Business (2001).

4 Dyson, E.: '*Digital identity management*,' Report in Release 1.0 (2002) — http://www.release1-0.com

5 '*Platform for privacy preferences (P3P) project*,' — http://www.w3.org/P3P/

6 Berthold, O. and Köhntopp, M.: '*Identity management based on P3P*', Dresden University of Technology — http://www.koehntopp.de/marit/pub/idmanage/p3p/

7 Godfrey, R. and Shaw, M.: '*Understanding the .Net My Services architecture*', PowerPoint presentation BSV304, .Net Systems Engineers, PDC2001 (2001).

8 Microsoft: '*Microsoft® .Net My Services, user experience overview*,' (October 2001) — http://www.microsoft.com/

9 Wong, W. and Lemos, R.: '*HailStorm promise and threat remain distant*', article in CNet News (30 August 2001) — http://news.com.com/2100-1001-272382.html?legacy=cnet

10 Montoya, M.: '*HailStorm May Be Dead, But Worries Persist*', Article in osOpinion.com (12 April 2002) — http://www.osopinion.com/perl/story/17251.html

11 Coursey, D.: '*MS: You don't trust us? OK, we'll open Passport, Hailstorm*', ZDNet (20 September 2002) — http://www.zdnet.com/anchordesk/stories/story/0,10738,2813501,00.html

12 Liberty Alliance homepage — http://www.projectliberty.org/

13 Data Protection Act 1998 — http://www.hmso.gov.uk/

14 European Directive 95/46/EC — http://europa.eu.int/comm/internal_ market/en/dataprot/law/index.htm

15 Apache Xindice native XML database — http://xml.apache.org/xindice/

6

SERVICE PERSONALISATION AND BEYOND

G De Zen

6.1 Introduction

The implementation of the concept of personalisation is traditionally based on the employment of user and service profiles. In addition, the mobile context requires the definition of a personalisation process, which takes into account the different adaptations and filtering that can be applied to content and services. Moreover, new requirements for service personalisation arise which impose, for example, the support of user profile modification and the handling of both static and dynamic profile information. The notion of location is the classical example of dynamic information that is used to personalise the offer of mobile services. But, many other concepts are becoming relevant like that of presence, environment and groups of users.

This chapter reports on the state-of-the-art as far as content negotiation, capability and preference profiles description standards is concerned and describes a proposal for a presence-aware personalisation system. Enhancements to mobile communications systems, in order to support presence-awareness concepts, are introduced. Collecting profile data, together with keeping it up to date with the changing needs and context of the user, are important issues.

6.2 Service Personalisation

Personalisation is recognised as being one of the most important factors in determining mobile service success. In fact, in the mobile environment, issues like service adaptation to terminal capabilities and to access network characteristics, and content filtering on the basis of user preferences, become even more relevant because they allow for prompt provision of services tailored to users' specific needs.

This chapter, which was first published as part of the Proceedings of FITCE Congress 2002, held in Geneva, is reproduced here by kind permission of the Federation of Telecommunications Engineers of the European Community (FITCE).

In order to identify which approach to service personalisation could be the most promising for future third-generation (3G) services, it is necessary to understand current trends and approaches to service personalisation and user profiling.

6.2.1 3GPP Reference Concepts

The 3GPP view about service prsonalisation is based on two concepts:

- personal service environment (PSE);
- virtual home environment (VHE).

The PSE [1] describes how users wish to manage and interact with their mobile services. It deals with subscription information (detailing provisioned services), preferences associated with those services, terminal interface preferences, and other information related to the user's experience of the system. Within the PSE the user can manage multiple subscriptions, for example both business and personal, multiple terminal types, and express location and temporal preferences.

The VHE concept [1] comes from 2G systems, namely the global system for mobile communications (GSM), and specifies an implementation of the concept of PSE that is focused on the portability of personalised information across network boundaries and between terminals. But, it is rather poor and based simply on the notion of profiles. The concept of VHE is such that users are consistently presented with the same personalised features either in the home or visited mobile network. It must be noted that VHE can rely on tool-kits like CAMEL (customised application mobile enhanced logic) [2] and OSA (open service architecture) [3] for implementing service personalisation.

The PSE contains personalised information defining how subscribed services are provided and presented towards the user. Each subscriber of the home environment (HE) has their own PSE. The PSE is defined in terms of one or more user profiles. The HE is responsible for providing and controlling the PSE of its subscribers.

The PSE allows for the personalisation of services offered directly by the HE (by the mobile operator) and by the value-added service providers (VASPs) who have an agreement with the HE for service provisioning.

As shown in Fig 6.1, we can distinguish two types of profiles:

- user profile, containing user-related data which is independent of services;
- user service profile, containing preferences associated with subscribed services provided by the HE and HE-VASP.

Normally, users cannot directly handle their profiles. But, this limitation is going to be overcome in Release 5 of Universal Mobile Telecommunications System (UMTS). In fact, 3GPP [1, 4] has introduced new requirements, which allow for user and HE-VASP access to profile data, and user self-configuration and management of services.

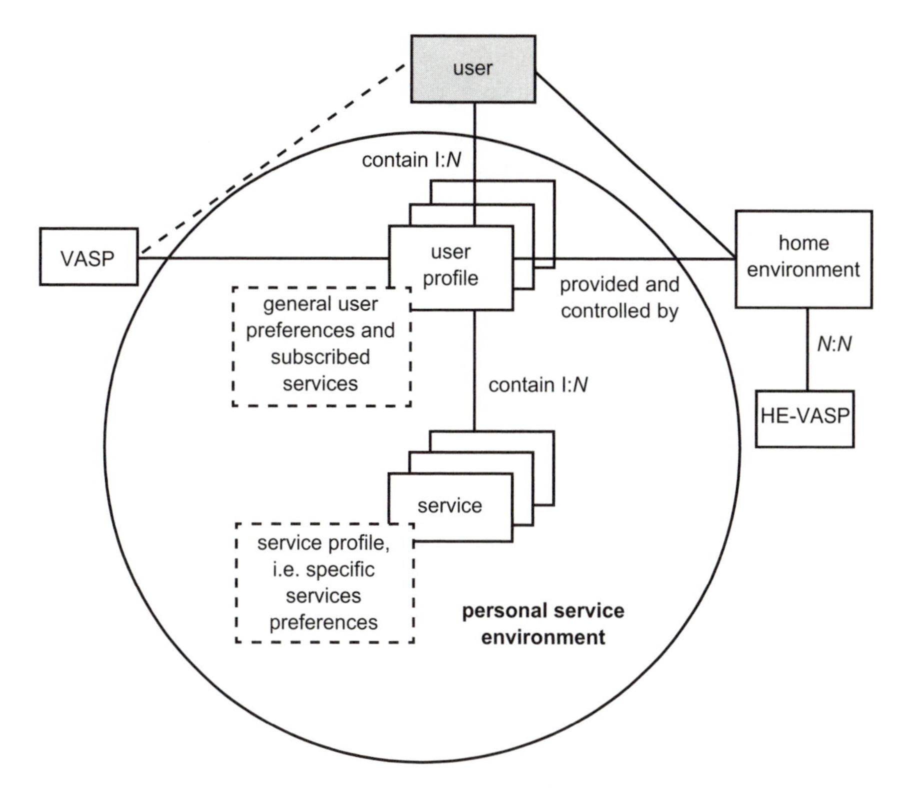

Fig 6.1 Personal service environment.

6.2.2 IP Multimedia Subsystem

The IP multimedia subsystem (IMS) [5] offers VHE concepts by supporting both CAMEL and OSA-based solutions.

CAMEL enables an operator to offer access to services even when the user is roaming by guaranteeing HE service personalisation. It should be noted that there is no requirement for any operator to support CAMEL services for their IMS subscribers or for inbound roaming users.

OSA allows for access to personalised services as well. In this scenario, the OSA service capability server is one of the three types of application server (AS) communicating with serving call session control function (S-CSCF) in the IMS.

In general, service personalisation is implemented by the application servers, which are contacted by the S-CSCF when an incoming or outgoing session initiation protocol (SIP) message matches one of the criteria contained in the user profile. This user profile is stored in the home subscriber server (HSS) and retrieved by the S-CSCF when the user registers.

6.2.3 Mobile Internet

Both in the mobile and in the fixed Internet, service personalisation is frequently implemented through portals [5] where user requests are processed and documents are dynamically generated or selected on the basis of user preferences. Such preferences are typically collected during user registration (the user is then provided with a username and a password). Alternatively, preferences are based on the history of visited links or on cookies exchanged between the browser and the server.

Sometimes, Internet services personalisation takes into account user terminal capabilities, i.e. browser capabilities. For this reason, the World Wide Web Consortium (W3C) in collaboration with the Wireless Application Protocol (WAP) Forum has defined a standard way for allowing browsers to inform Web servers about client capabilities and preferences, namely composite capabilities/preference profiles (CC/PP) [6]. The motivation for this standard is the need to handle in the future a large variety of Web terminals (e.g. laptops, PADs, smartphones) having different hardware and software platforms, browser capabilities and service support features.

The CC/PP profile, which matches the user agent profile (UAProf) [7] defined by the WAP Forum, is included or referenced in the HTTP requests sent by the client.

Using an extended capabilities description, an optimised presentation can be produced. This can take place by selecting a style sheet that is transmitted to the client, or by selecting a style sheet that is used for transformations. It can also take place through the generation of content, or transformation, e.g. using extensible style-sheet language transformations (XSLT).

The use of CC/PP enables service personalisation based on content negotiation.

6.2.4 WAP Forum

As introduced in the previous section, the WAP Forum has addressed the problem of service personalisation in terms of content negotiation issues. Clearly, the need for delivering contents fitting mobile terminals' capabilities and for saving radio resources has driven the WAP Forum to adhere to the CC/PP approach and to identify the components which are useful for describing the user agent; that is, the terminal. In effect, the user agent profile is composed by named components each containing a collection of attribute-value pairs.

As stated in the UAProf specification [7], classes of device capability and preference information are identified. The classes include the hardware and software characteristics of the device as well as information about the network to which the device is connected. It must be noted that UAProf does not contain any application-specific information about the user.

6.2.5 SIP

The Internet Engineering Task Force (IETF) has two solutions which enable SIP-based services personalisation.

The first is the call processing language (CPL) [8]. CPL is a language that has been designed by the IPTEL working group for describing and controlling Internet telephony services. It allows a SIP client to include, within a SIP message (typically Register and Invite), call processing instructions for incoming and/or outgoing calls that have to be processed by SIP servers. The CPL is, at the same time, a simple and a powerful language. It is designed to be easily created and edited by the user but it is also aimed at describing a large number of services and features. However, it has been conceived so that there is no way for possible misuse by users. One of the key characteristics of CPL is that it is based on the extensible markup language (XML). CPL can be used for configuring advanced call services like time-of-day routing, outgoing call screening, priority and language routing, call forwarding, etc.

The second solution is represented by the 'SIP Caller Preferences and Callee Capabilities' Internet-draft [9] which has been proposed to solve the problem of incorporating the sender (caller) preferences to SIP messages to enforce specific routing decisions, e.g. the caller wants to contact a user but not their voice mail. These preferences allow an indication of where to proxy or redirect a request and which request handling directives are needed. For supporting this, three new request headers have been defined. The draft also proposes to extend the contact header to handle new parameters that can be used for describing both user and user-agent (UA) characteristics, like supported media, spoken languages, priority, type of terminal, etc.

6.3 User Profiling

6.3.1 User Agent Profile

As introduced in the previous section UAProf is used for service personalisation and, in particular, for content adaptation.

The user agent profile schema is based on the resource description framework (RDF) schema and vocabulary as foreseen by CC/PP. To date, the following components have been defined:

- hardware platform;
- software platform;
- network characteristics;
- browser UA;

- WAP characteristics;
- push characteristics;
- MMS characteristics.

All the attributes defined for these components refer to static characteristics of the mobile terminal hardware and software platform except for the current bearer service attribute of the network characteristic component that indicates the bearer on which the current session is opened.

6.3.2 Generic User Profile

The generic user profile (GUP) represents a very interesting solution to the problem of defining a flexible and extensible user profile which all mobile network actors can refer to univocally.

It is currently under standardisation by 3GPP for Release 6 [10]. Among the requirements GUP has been designed to fulfil, some of the more interesting are:

- to deal with network distribution of user related data which causes difficulties for users, subscribers, network operators and service providers to create, access and manage data located in different network entities;
- to provide a reference conceptual description to enable harmonised usage of the user-related information regardless of actual data location (for example, user equipment, home network, serving network and VASP equipment);
- to allow extensibility;
- to support user profile access and management by different stakeholders such as the user, subscriber, service provider and network operator by a standardised access mechanism;
- to enable the user to handle subscribed services differently on the basis of different usage contexts.

As defined in the GUP architecture [11], the user profile data description is handled by identifying the following domains (see Fig 6.2):

- data — data stored and/or accessed in a user profile;
- data description — describes the data contained in the user profile (this is also called the schema level);
- data description framework — defines how to create the data description (CCPP-RDF, XML schema).

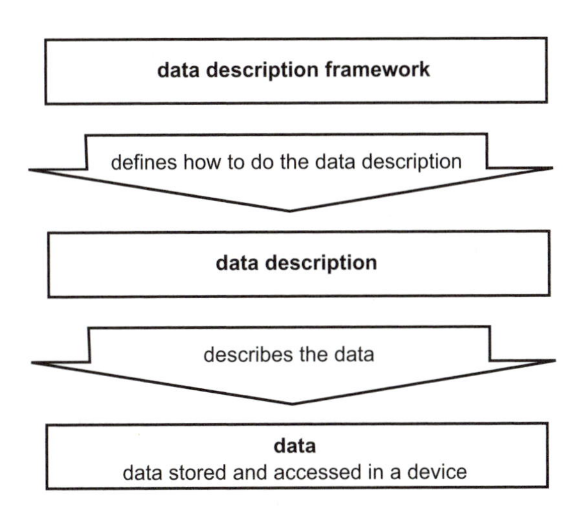

Fig 6.2 Data description framework logical levels.

The generic user profile consists of user profile components which allow for logical grouping of related data. A user may have one or more instances (specific values) of a specific user profile component. Only one instant of a specific component is active in a specific situation. When the GUP data is used or accessed, only a subset of all the component instances is relevant, i.e. active. This subset identifies the active profile. A classification criteria of data in GUP components is based on information characteristics [10]:

- general information:

  — general user information (name, address, age, sex, ID);

  — logical identifiers (e.g. logical name, personal number, e-mail address);

  — general subscriber information (name, bill information, users);

  — general privacy preferences;

- capability description:

  — terminal capabilities (e.g. user interface capabilities, communication capabilities, synchronisation capabilities, MExE capabilities, WAP browser capabilities);

  — subscribed network capabilities (the subscriber will be allowed to gain access to the set of subscribed capabilities);

— roamed-to network capability (the user will be allowed to gain access to the set of subscribed and supported network capabilities);

— subscribed service capabilities;

- user's preferences:

  to allow users to describe their wishes and to indicate preferences for specific contents or information or policy, e.g. browser appearance, preferred memory usage;

- service customisation (to customise subscribed user services or applications):

  — user interface (ring volume, ring signals, melodies, key sound);

  — WAP parameters (bookmarks, gateway, Internet account, gateway IP address, user ID, password, data mode, security, show images, response timer);

  — user security policy (application download, ciphering, positioning);

  — user security data (secret keys, user name);

  — authentication data (e.g. password, pin, voiceprint);

  — supplementary services settings;

  — quality of service associated to the user;

  — status of services (active/not active).

6.4 Presence Service

Implementation of the personalisation concept is traditionally based on such static user and service profiles. However, new requirements for service personalisation have arisen which require the handling of such dynamic information too. Examples of such dynamic user-related information are:

- connection status (on-line/off-line);
- geographical user location (e.g. country, city);
- user context (e.g. home, office, vacation);
- willingness to communicate (e.g. available, busy);
- preferred media for communication (e.g. voice, instant messaging, e-mail);
- call state (e.g. busy, free).

Such dynamic information is referred to as presence information and in general can be characterised as either user or network provided. The presence concept comes from the Internet world where it has been demonstrated to be very successful for supporting user interaction via message exchanges. In fact, instant messaging applications, like AOL, Yahoo, Microsoft Messengers and ICQ, have become very popular on the Internet and are now expected to be used on mobile telephones too.

3GPP has recognised the relevance of presence as a service-enabling technology and it is going to standardise an enhanced presence service integrated in the mobile network [12] that can be used for customising a variety of services. The presence service defined by 3GPP is aligned with IETF proposals, i.e. common presence and instant messaging (CPIM) [13], and SIP for instant messaging and presence leveraging extensions (SIMPLE) [14], which provide solutions to existing applications interoperability issues.

6.5 Service Personalisation Process

With respect to the problem of supporting personalisation of 3G multimedia services, Siemens approach is that of defining a GUP-based profiling system and a personalisation process which both leverage on it. The envisaged solution allows for implementing a PSE able to satisfy the following mobile user requirements:

- user self-provisioning via a graphical user interface;
- user configuration and management of active profile;
- service personalisation based on both static and dynamic user-related data.

It is expected that service providers, content providers and network operators will collaborate to offer personalised services to mobile users in a way that satisfy user's individual preferences and needs at a specific time and place.

The proposed personalisation process (see Fig 6.3), goes through the following steps:

- contents filtering on the basis of presence information if required by the service and authorised by the user;
- contents selection on the basis of:

  — either static service customisation preferences or generic user preferences;

  — dynamic service customisation preferences provided with the service request (these preferences may override the previous);
- contents aggregation;
- content adaptation on the basis of:

  — user provided presence information;

  — network provided presence information;
- terminal adaptation on the basis of user-provided terminal capabilities information.

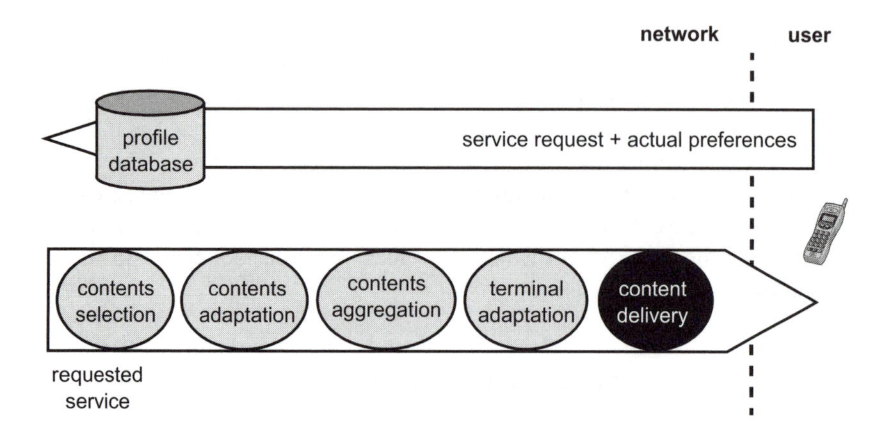

Fig 6.3 Service personalisation process.

A programmable rule-based system can be used in order to allow the user to define a personal personalisation process.

At the standardisation level, issues related to the enrichment of mobile service architecture in order to enable service personalisation are beginning to be tackled and this will be one of the areas of major activity for future fourth-generation systems, as also reported elsewhere [15].

6.6 Summary

This chapter has looked at the state-of-the-art as far as the Internet and the mobile world solutions for service personalisation and user profiling are concerned. The main characteristics of existing solutions have been identified and explained.

In addition, a proposed approach to service personalisation has been described. The solution is based on an enhanced profiling system which follows the GUP paradigm and a personalisation process which exploits both user profiles and presence information.

References

1 3GPP TS22.121: '*Service Aspects, The Virtual Home Environment*', Release 5 (2002).

2 3GPP TS23.127: '*Virtual Home Environment/Open Service Architecture*', Release 5 (2002).

3 3GPP TS 22.078: '*Customised Applications for Mobile Network Enhanced Logic*', Release 5 (2002).

4 3GPP TS23.228: '*IP Multimedia Subsystem (IMS)*', Release 5 (2002).

5 Mandata, D., Kovacs, E., Hohl, F. and Amir-Alikhani, H.: '*CAMP: a context aware mobile portal*', IEEE Communications Magazine, **40**(1), (January 2002).

6 W3C CC/PP — http://www.w3c.org/

7 WAP-248-UAPROF-2001 1020-a: '*User Agent Profile*', (2001).

8 IETF Draft: '*CPL — a language for user control of Internet telephony services*', draft-ietf-iptel-cpl-06.txt (2002) — http://www.ietf.org/

9 IETF Draft: '*Caller preferences and callee capabilities for SIP*', draft-ietf-sip-callerprefs-08.txt (2003) — http://www.ietf.org/

10 3GPP TS 22.240: '*Service Aspects, Service Requirement for the 3GPP Generic User Profile (GUP)*', Release 6 (2002).

11 3GPP TS 23.240: '*3GPP Generic User Profile — Architecture*', Release 6 (2003).

12 3GPP TR 23.841: '*Presence Service Architecture*', Release 6 (2002).

13 IETF RFC 2779: '*Instant Messaging/Presence Protocol Requirements*', (2000) — http://www.ietf.org/

14 IETF Draft: '*A presence event package for SIP*', draft-ietf-simple-presence-10.txt (2003) — http://www.ietf.org/

15 Wireless World Research Forum: '*Book of Visions*', (2001) — http://www.wireless-world-research.org/

7

PROFILES — ANALYSIS AND BEHAVIOUR

M Crossley, N J Kings and J R Scott

7.1 Introduction

Recent research [1] indicates that over 30% of the knowledge worker's day is involved in searching for required information.

The amount of raw information available is growing at a phenomenal rate. Lyman and Varian [2] indicate an annual production of 1.5 billion gigabytes of information. This growth buries the relevant information that the user requires under an avalanche of reports, e-mails, documents, and information from portals and Web sites.

Information overload has long been recognised as a significant issue in the workplace [3, 4], with increasing numbers of senior executives believing that the situation is getting worse. Conversely these senior managers also believe that, as an asset, their company's information is mission critical.

Among the many definitions of knowledge management, the simplest is often phrased along the lines of:

> "... getting the right information to the right people at the right time".

Behind this definition are two inherent questions which must be answered:

- What is relevant [information] to the user right now?
- Who is relevant [people] to the user right now?

Where once people used to have tightly defined roles in their company or within their project, it is becoming increasingly common for people to take on multiple technical and managerial roles. For instance, an individual could spend the morning working as a project leader on one project, but the afternoon as a technical expert on a separate project. What constitutes the right information is thus becoming increasingly related to the user's current context. Yet systems typically do not understand the user's context — information is provided even if it is not relevant to what the user is doing at the time.

To address these issues, two further fundamental problems need to be tackled:

- understanding content — what a particular document is all about;
- understanding interests — what the user is interested in, both historically and in the current context.

The concepts underlying personalisation are introduced in Chapter 1. Building on earlier work within BT, this chapter presents work done to help deliver the right information to the right people at the right time, based on their profile, both static and dynamic. Figure 7.1 shows the range of profiling which can be considered.

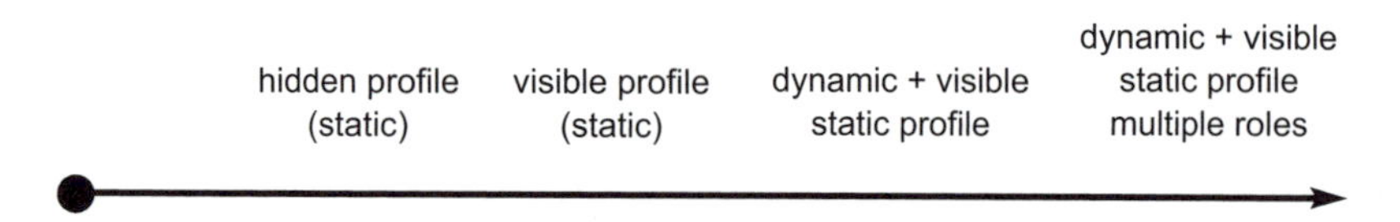

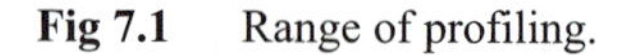

Fig 7.1 Range of profiling.

Section 7.2 describes existing knowledge-management tools, in particular how they are personalised using the user's profile.

Section 7.3 investigates how user profiles can be created, and how different aspects of profiles can be used.

In section 7.4, this chapter describes a new method for understanding the user's focus (current activity), and creating a dynamic profile based on this current activity.

Using the user's created dynamic profile, section 7.5 shows how the dynamic profile can be used to improve the relevance of information delivered to users.

7.2 History

BT has a rich history of developing a number of highly successful knowledge-management tools going back several years. Many of these tools already use personalisation to improve the quality of information delivered to the end user.

7.2.1 Knowledge-management tools

A number of BT's knowledge-management tools and research projects are described briefly below, with particular emphasis given to their use of profiling.

7.2.1.1 Document Summarisation

ViewSum is a text-summarisation tool that can provide a personalised summary of any document. Depending on the user's needs, it can summarise the document to any required length — even to a single sentence or set of keywords. It can also exploit the user's profile to tailor the summary to the individual user.

ProSum is the text-summarisation software component from which ViewSum is built, and which has been incorporated into a number of other BT knowledge-management tools.

7.2.1.2 Document Retrieval

ProSearch [5] software agents find and filter relevant information that matches a user's interests. ProSearch can search the World Wide Web, a company intranet, news groups, on-line news sources and more.

As such it will search these sources based on the user's profile and then filter the results by removing duplicates, documents already seen by the user, and documents that are referenced but no longer exist, thus overcoming a number of user frustrations with 'traditional' search engine technology.

Consequently ProSearch presents a set of relevant, checked documents that the user has not yet read.

7.2.1.3 Document Search Engines

Finder, QuizXML and QuizRDF [6] are search engines for the Web, XML document sets, and RDF-annotated document sets respectively. There are currently no personalisation aspects to these search engines.

7.2.1.4 Document Sharing

Jasper [7] is a collaboration tool that provides an information repository and discussion forum for communities of practice. Documents shared within Jasper are matched against members of the community, and only relevant users are notified. In addition, the Jasper agents learn more about the user's profile as the user recommends documents to Jasper or provides feedback on other submitted documents (see Fig 7.2).

Further details of Jasper's profiling techniques are discussed in section 7.4.1.

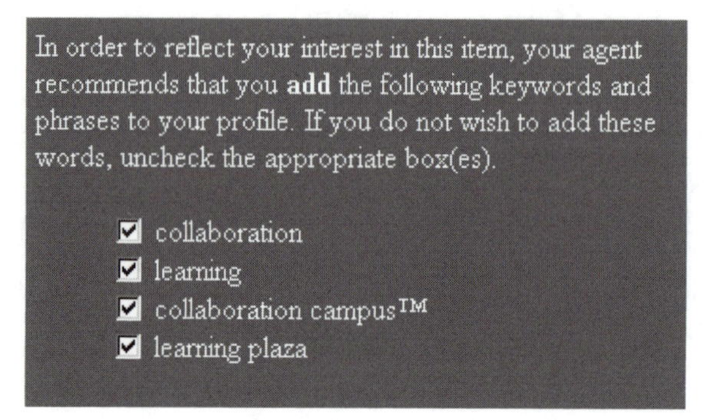

Fig 7.2 User profile feedback within Jasper.

7.2.1.5 Document Visualisation

The Knowledge Garden [8] is a collaborative 3-D information visualisation tool for finding, retrieving and sharing information. The garden metaphor reflects the organic, changing nature of information. Plants represent information resources, and can grow in the garden based on a user's personal interests, or the interests of a shared community (such as those from the Jasper system).

In the Knowledge Garden shown in Fig 7.3, Jasper documents are represented by the heavily populated sector, which is seen as a common area by all users of the Knowledge Garden. The sector immediately clockwise from the Jasper sector could be employed by the user to represent their personal documents (for instance, a plant could represent their browser bookmarks). This sector is thus personalised to the user, and is not visible to other users unless the owner grants access.

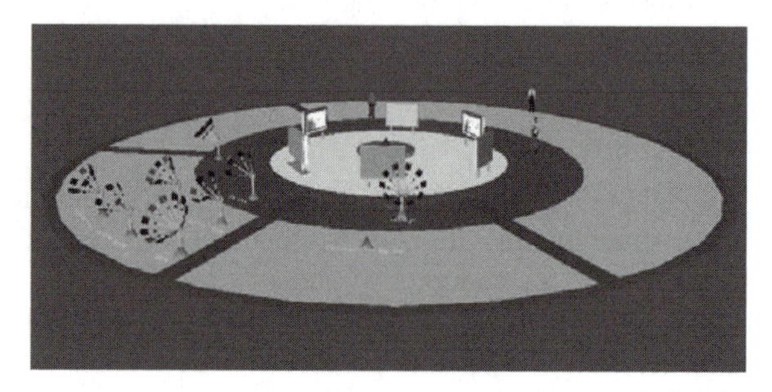

Fig 7.3 Knowledge Garden.

Personalisation is often at odds with the basic concepts of shared spaces. It is technically feasible for plants to be arranged based on the user's profile (for instance, more important plants to be placed close to the garden entrance). However, this can cause interface difficulties when multiple users see different spatial arrangements as they navigate the same shared space.

7.2.1.6 Shared Meeting Spaces

A 'community of practice' [9] describes the informal groups where much knowledge sharing and learning takes place over a period of time. Essentially it is a group of people who are peers in the execution of real work. Therefore, the community of practice is not typically a formal team, but an informal network, each sharing in part a common agenda and shared interests.

The Forum, which consists of the Contact Space, Meeting Space, and Jasper components, was designed to facilitate communication and information-sharing where there are mutual concerns or interests among virtual communities of practice within or across organisations [10]. In the early development of the Forum, it was hypothesised that in large, geographically distributed organisations, such as BT, it would be very difficult to know everyone who could be in one's community of practice without the assistance of a community-based system. Therefore, the Forum was designed to serve as a virtual meeting space where strangers with common professional interests could participate in work-related topics and come together in the virtual space to get to know others in the organisation [11].

The Forum provides user profile matching, collaborative information storage and retrieval, summarisation of shared documents, graphically enhanced audio-conferencing, and a dynamic 3-D virtual world that moves users with common tasks together in the 3-D environment with the goal of supporting spontaneous chat interactions. Thus, the forum was developed with the express purpose of supporting informal communication and information-sharing among members of a community of practice.

7.2.1.7 Personalised Portals

The RIO portal prototype (an internal BT research project) provides a personalised portal, where users can build the portal themselves based on their interests. Using their personal preferences, they can change the screen layout and resources viewed. News stories and personalised summaries of news items are presented based on the user profile. Users with similar interests can be identified.

7.3 The Use of Profiling

A profile can be considered as capturing sufficient information to model certain aspects of a user's behaviour.

7.3.1 Sources of Information

In order to build user profiles, data about the user must be collected. There are a number of methods available for doing this, including the following.

- Data-mining user's directories

  A prototype Java tool has been developed with which the user can generate a 'first cut' of a user profile. The Java tool is pointed at a user directory (typically the PC's 'My Documents' folder). The tool then uses ProSum to extract keywords from all document files, aggregates the keywords across all these document files, and gives a first approximation of the user's profile. Further facilities could be added, such as finding all Word documents authored by the user (from the Word document properties), and giving greater weighting to keywords found in these documents. However, such techniques rely on the appropriate metadata being correct; this is often not the case.

- Data-mining Web-server log files

  This method has an issue surrounding data validity as the user may access Web pages via a proxy server. A user may look at a Web site's pages with some frequency but those pages may be cached in either the local client cache or a network proxy server. Therefore the total hits by that user would not be accurately recorded on the site. Another major problem here is that since each Web server keeps its own server log, and users move rapidly between sites, it proves very difficult to create a full and consistent profile of a user.

- Cookies

  A cookie is a data file that is placed on a user's local disk by a remote Web server they have visited. It is then used to uniquely identify a user during Web interactions with that particular site. It enables the remote Web server to keep a record of who the user is and what actions they take at the remote site. The contents of a cookie will vary according to the site, but typically they can store the following information — personal ID, recent activities at the Web site, site-password information, details of items purchased. They may also be used for marketing by allowing monitoring of interest to particular advertisements, services or pages on a Web site. However, it should be noted that a user can set their browser to disable cookies.

- CGI identification [12]

  This system uses CGI scripts to track someone through a site. It is used where URLs have long strings containing a user ID and a session ID which are the internal identifiers passed from page to page in the server. When a script is called, it looks up the values of the variables and incorporates them in calls to other pages. A drawback with this method is that it specifically requires the user to log in to identify them. However, a CGI application combined with the use of cookies could be used as a basis for a tracking process.

- Proxy servers

  Instead of directly accessing Web pages, a browser may send all of the page requests through a proxy server. Additional software could be placed within the proxy to extract information about the user requesting the page and the page itself.

- A knowledge service such as Engage Knowledge [13]

  This service employs the principle of having a central repository to hold user data and to which Web site providers subscribe, with data flowing both to and from the Web site and the repository. As more Web sites enrol in the service, a much richer picture about a user's behaviour can be built up. For instance, if a user is anonymous on one site, it is possible to infer their name, if they had been registered on a different site.

- Profile hosting server

  A more powerful method of central profiling is the profile hosting server, as described in Chapter 5. This allows an individual user to have finer control on the visibility and publication of aspects of their profile.

- Indicators and user dwell

  By using a client-based application, it is possible to observe and analyse user behaviours by monitoring indicators such as click-stream (the links that users click) and user dwell (the length of time a user spends on a given page). Through this analysis it is possible to infer user preferences and information requirements. [14].

7.3.2 Aspects of Profiling

This section deals with aspects of profiling and their associated information. Each aspect is discussed although some are covered more rigorously, as by their nature they are broader in scope. A number of the different aspects of profiling can be considered as different axes on a graph, as shown in Fig 7.4.

7.3.2.1 Declared/Inferred

Information gathered about on-line users falls into two basic types, that which is declared and that which is inferred.

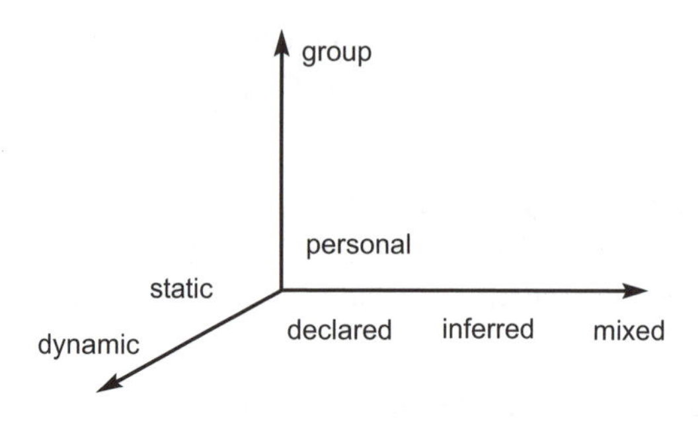

Fig 7.4 Aspects of profiling.

- Declared information

  This is information typically obtained through users filling out forms and providing identifiers such as name, e-mail address, post code, interests and ratings.

- Inferred information

  This is information that can be indirectly associated with users such as by identifying similar likes or interests, i.e. because X is interested in Y it is inferred that they will also be interested in Z. For example, if a person has stated they have an interest in Ipswich Town Football Club it may be inferred that they would also be interested in any rule changes announced by the Football Association that may affect the club.

Declared and inferred are also referred to as active and passive profiling techniques respectively by some people working in the personalisation domain [14]. Also, the declared data can be referred to as explicit information while inferred can be referred to as implicit information.

One issue related to declared profile information is that of user resistance; if a user is asked for too much information this can be a disincentive, for example, to remain on a given Web site or register for a particular service. Thus a scheme of progressive declared data capture needs to be employed [15]. For example, a system could be employed where on the tenth registered visit the system prompts for demographic data if it has not already been entered (such as when a user purchased something).

Avila and Sherwin [16] declare a third type of information called ‘behavioural information’. This is information that is passively recorded through user logins, cookies, and/or server logs. From monitoring the links a user clicks on, or the amount of time they are active on a Web page, it may be possible to gain an indication of a user’s interest in that item. However, simply monitoring the links a user has visited is not a reliable mechanism for deciding what a user’s interests may be.

Reliability can be improved by adding an explicit mechanism, where a user explicitly rates the 'desirability' of a page. Also, threshold settings, for read times, may be set by analysing the length of time it takes for a user to read a document (or, more accurately, the time a user's Web browser is displaying a given page), such as a job advert [14]. In addition to the read time, the number of revisits could be monitored from server logs.

7.3.2.2 Public/Private

With declared information a user has the ability to define profile elements as public or private with respect to other system users. For instance, they may wish to keep their date of birth private while allowing their name to be public. Indeed the system being used may impose a scheme where certain profile information, such as a user's name, must be declared public in order for the system to carry out its tasks effectively. An example of this would be where the system has a 'people like me' function which informs individual users of others who have similar interests.

There could also be varying degrees of 'public'. It could mean any Web users, users of a particular system, or users within the same community of interest. A clear definition of 'public' should be available to users in any system that differentiates between public and private. Greenspun [15], when describing the ArsDigita Community System, suggests a privacy attribute associated with each profile element that identifies it to be 'show to everybody', 'show to other registered users' or 'show only to maintainers'. The e-mail privacy attribute is set by default to 'show only to other registered users' to prevent SPAM crawlers from harvesting e-mail addresses from public Web pages.

7.3.2.3 Static/Dynamic

Profile information may be either static or dynamic. Static could refer to information that either never changes or changes very rarely — someone's date of birth will not change but their name may change upon marriage. People may have a set of core interests that do not change over their life and are therefore static. Or it could be that once a profile has been set up upon creation of a user account, a user does not make the effort to maintain that profile. Other profile information will be dynamic, constantly changing, as a user's needs change. Their interests may change or expand or their level of knowledge/skill level grow leading to a change in the type of content in which they are interested.

Making the profile easy for a user to update is crucial, with the ability to incrementally add or change interests. However, any tool must still be able to work correctly with the minimum of information to allow a user to get started simply, without supplying an excessive amount of detail. An additional requirement would be to apply a greater importance to terms added most recently to a profile. Within ProSearch, for example, a user is able to express a simple starting profile for a Web search, but then tune that search to be more complex, based upon feedback on returned pages.

A person's profile could be inferred from a wide range of sources, such as the content of documents held on their local PC, a person's e-mails and even their contributions made to discussion groups. In that case the amount of information gathered can become excessive, so boundaries need to be set. Based on their experience, the ArsDigita Community System [15] limits the storage of data to only that from registered users.

7.3.2.4 Updating the User's Profile

Profiles are not static, and provision must be made for users to change their profile as their interests change over time. With manual methods of updating profiles, it is notoriously difficult to ensure that users keep their profiles up to date.

Early versions of Jasper would automatically update a user's profile based on their feedback. This technique is also used by a number of other commercially available knowledge-management tools.

In terms of user acceptance, a better method, as shown by user feedback from early Jasper trials, is for a system to work in a semi-automatic way, and prompt for user confirmation of any inferences made. Within Jasper and ProSearch, the system is able to infer which keywords to add to, or remove from, a profile, but those choices are always displayed for user confirmation before the profile is updated.

7.3.2.5 Roles

A user may adopt different roles. For example a person's role typically changes between work and home. This could lead to a change of personalised home page/ personalised portal accordingly. Users could either declare which role they are in or the role change be inferred on a time of day basis with the ability to switch roles manually if required. In the workplace, this would be desirable where a person undertakes a number of distinct roles, such as project manager for one project or team member of another. The information that appears on their personalised page could then change accordingly, along with their access rights to information.

It could also be that a person has more than one profile. For example they could have different profiles according to the mode they are in such as work, home or travel. Obviously this could also just be a case of utilising particular aspects of one larger profile dependent upon the mode they are in and the role they have within that mode.

7.3.3 Personal and Non-Personal Information

In designing a system to hold personal information, there must be a thorough understanding of what information may be published or circulated about a user. Some information must be treated with greater sensitivity, such as medical records, where a medical professional or a health service administrator may have rights to receive some relevant information, but not the complete contents of a given profile. Information such as a password is quite obviously personal as is any password cue that is used in a system for when a user forgets their password. It is imperative here to consider cultural differences. For example, the use of a mother's maiden name as a password cue is not acceptable in some countries such as Iceland as it is easily deduced from a person's surname.

7.3.3.1 Personal/Group

When considering the relationship between groups or communities Fernbad and Thompson [17] state:

> 'The notion of community is a 'public' concept in that it entails a collectivity of sorts. But virtual community has a private quality about it; it may be who we are as private individuals that constitutes our membership in certain communities, e.g. virtual communities based on political ties or communities of interest based on world view, hobbies or professional status. Thus a private character is ascribed to the idea of a community as our individuality increasingly defines our choice of community membership, despite the nature of community as a social bond.'

This emphasises the relationship between the users and groups of which they become members, and thus suggests a relationship between a user and group profile. It would be expected that there may be duplication of profile data between the two. This could be interests, bookmarks and popular pages.

7.3.3.2 Group Profiles

When groups or communities are being supported electronically, one of the first sets of tools provided is usually the ability to share files, such as documents or bookmarks. A common approach to sharing URLs is by means of server-based bookmarks that focus on a group's interest. Although not specifically related to group bookmarks, Herzberg and Ravid [18] identify the problems of such server-based indexes and some of these also relate to groups. These problems include:

- dependence on the index server availability;
- slower access than locally held bookmarks;
- the same bookmark may be relevant to more than one folder;
- different users may prefer a different arrangement of folders;
- users often have private bookmarks they do not wish to share.

Their suggested solution uses the following ideas:

- common taxonomy of attributes, customised trees of folders — the community all use the same set of attributes (taxonomy);
- each user can build their own tree of folders, assigning attributes to each folder;
- each bookmark is also assigned attributes, which are used to map the bookmark into the appropriate folders for each user;
- the database of all bookmarks is kept in a server to which all users have access and local replicas of this database are kept on each user's machine — this replication enables users to work off-line and provides better performance;
- users may easily define some URLs (and possibly attributes) as private — privately marked URLs are encrypted in the server and in replicas of that user, so that access is only possible using the key of the user.

With a shared bookmark system, users could add relevant URLs to group bookmarks, add a weighting to them and identify themselves as the bookmark author to allow others to see who added it. Note that some trials [15] have found that users have been more inclined to comment on a site rather than add a rating about it, despite the fact that it was easier and faster to type a rating of '5' than to add a comment.

One implementation embodying these ideas is Jasper where users are notified via e-mail of new URLs added to their interest group. In this case not all group members will necessarily receive notification, as Jasper also evaluates the URL with respect to their personal profile settings.

Demographic information may also be used to personalise the customer experience as is demonstrated by the following from BroadVision [19], an eCommerce software company:

> "Customer personalisation involves defining communities of Web site visitors with similar demographics, spending power and interests. When potential customers log on, they see content tailored to their preferences or budget. Personalisation works in another way through the Infoexchange Portal, which considers the user's role, location, objective and access device, and provides them with appropriate information."

7.4 Understanding Focus

Not only do people have many and varied interests, they also have differing interests at different times (their current interests being their 'focus'). A key to solving information overload is to understand this focus, and deliver information to the user which is relevant to their current activity. A related issue is that the information delivery should not disturb the user's normal activity; feedback must be timely and appropriate.

Jasper has been used within BT for a number of years, providing a static analysis of a user's profile. The recently developed Peruse system, however, takes the profile a significant stage forward by discovering and utilising the user's current context.

7.4.1 Jasper Profiles

As mentioned briefly above, Jasper is a Web-based collaborative virtual environment comprised of a system of intelligent software agents that hold details of the interests of their users in the form of user profiles [20]. Jasper agents summarise and extract key words from World Wide Web pages and other sources of information (e.g. personal notes, intranet pages, applications) and then share this information with users in a virtual community of practice whose profiles indicate similar interests. Jasper has been used by a number of geographically dispersed communities, inside and outside BT. Typical Jasper communities have thirty members, but the largest community to date has over three hundred registered participants.

Jasper agents store, retrieve, and summarise information. Each Jasper user has a personal agent that holds their profile based on a set of key phrases which models their information needs and interests. Thus, Jasper agents modify a user's profile based on their usage of the system, seeking to refine the profile to better model the user's interests.

In order for a user to share information with the community of practice, a 'share' request is sent to one's Jasper agent via a menu option on the individual's WWW browser. Jasper extracts the important details from these documents, including the summary and keywords, as well as the document title, its URL, and the date and time of access. Users may annotate the document, and specify one of a predefined set of interest groups to which to post the information being stored.

This information captured by the Jasper agents is maintained locally and serves several purposes. Firstly, the summary provides the user with an abridgement of the document without the need to retrieve the remote pages unless greater detail is desired at the time. The ProSum text summariser extracts key theme sentences from the document based on the frequency of words and phrases within a document, using a technique based on lexical cohesion. Secondly, the content of the HTML page is analysed and matched against every user's profile in the community of practice. If the profile and document match strongly enough, the Jasper agent e-mails users, informing them of the page that has been shared, and by whom. Thirdly, the posted information is also matched against the sharer's own profile. If one's profile does not match the information being shared, the agent will suggest phrases that the user may elect to add to their profile. Thus Jasper agents adaptively learn their users' interests by observing their users' behaviour.

Jasper agents contribute to the development of the shared repertoire of a community of practice by providing a repository tool for managing information and sharing resources. Jasper aids the community in co-creating their repertoire by providing the virtual place where users can manage information in an historical context (i.e. who posted what, when, why, and with what effect).

The user information stored within Jasper forms the basis for the user's long-term, static profile. Each user profile contains a list of words and phrases, which have been identified as useful to that particular user. In addition to that list, each person can subscribe to a number of interest groups within the community of practice. Through feedback given to a user's Jasper agent, the profile is adapted to reflect one's current set of interests. By analysis of usage patterns, Jasper is able to identify the latest topic or the topic of most interest to the community.

However, this static analysis is insufficient to support and facilitate a culture of innovation. Brown and Duguid [21, 22] suggest that the opportunistic movement of a document through an organisation is almost as important as the document itself, i.e. the context in which a document is used creates a socially negotiated meaning. Much of the context of an organisation is communicated through informal channels. Although users of Jasper could post and retrieve information, this was often not enough to foster informal communication between users, especially if they are strangers. Jasper usage patterns indicated that notably few users ever decided to e-mail someone they did not know. Additionally, the Jasper repository provides no opportunity for community members to interact synchronously [23].

7.4.2 Peruse

In order to facilitate interaction between users, a more dynamic method of delivering information based on the user's current context was designed. The system developed, Peruse, consists of a network of PC-based client applications, with a centralised classification server:

- the PC client monitors Web browsing activities, and activation of certain PC applications and, as a consequence, messages are sent to the central server detailing user events, utilising the HTTP protocol — at appropriate times, the client will request Web page suggestions for the user to visit;
- the central server classifies the Web pages being browsed, and identifies additional resources that are relevant to the user's current focus.

The user will be sent information as to who shares interests, or are in the current interest cluster, in order to facilitate social networking.

In any given community, someone may already have found the information a user requires, whether this requirement is business or socially related. Peruse is intended to support a community of people in gathering information, as well as supporting repetitive activities:

- by publishing, in a safe manner, the activities of the members of a community, this will build up and re-inforce the sense of community identity;
- by observing the browsing activities of a community, Peruse is supporting the notion of on-line 'serendipity', where new information sources are placed into the awareness of a user, without forcing the issue.

7.4.2.1 *Client Functionality*

The client monitors the focus of a user's activities on their PC. Messages are sent to the central server, containing information such as URL of Web page visited, time spent reading the Web page, and number of unread e-mails. For each Web page visited, a number of messages are generated — when the page is first visited, when the page is exited (with cumulative visit time), and when the page is brought into and out of focus.

Based upon the local use of the computer, the client identifies 'boundaries' or changes in work patterns, which are appropriate times for the user to be notified about additional Web sites.

Currently, this is implemented by a request for information being sent when the screen saver is switched off, i.e. when the user returns to the computer. However, further information, such as speed of keyboard activity, calendar bookings, and task list entries, could be included to give a more detailed picture of a user's focus of attention.

7.4.2.2 *Server Functionality*

The server determines the current activity focus, in the following manner. The description describes how Web pages are analysed; however, the principle can be extended to any form of electronic media.

A Web page (URL) is received from the client, for a particular user. The server fetches the Web page, and keywords are identified and extracted from that page. ProSum is used to extract those significant keywords from the page, though other mechanisms such as keyword frequency could be utilised. For each keyword, or phrase, found, a record is made of the Web page (URL) where the keyword was found. Hence, a record is made that a particular user has visited a particular URL;

the keywords from the current page are weighted, and are then added to the user's dynamic profile.

Weightings are given to each keyword (as shown in Fig 7.5) extracted from that Web page, based upon a modified Rayleigh function [24]:

$$\beta = \frac{E\sqrt{2}}{\sqrt{\pi}}$$

$$R(t) = \frac{EKt}{\beta} e^{(-\frac{1}{2}(t/\beta)^2)}$$

The weighting added to each keyword is $W = R(t)$, where t is the time a particular Web page was in focus (i.e. was the currently selected window). E is the calculated mean viewing time (average length of time any page is viewed), for a particular user, and K is a scaling constant, which is currently set at a value of 5.

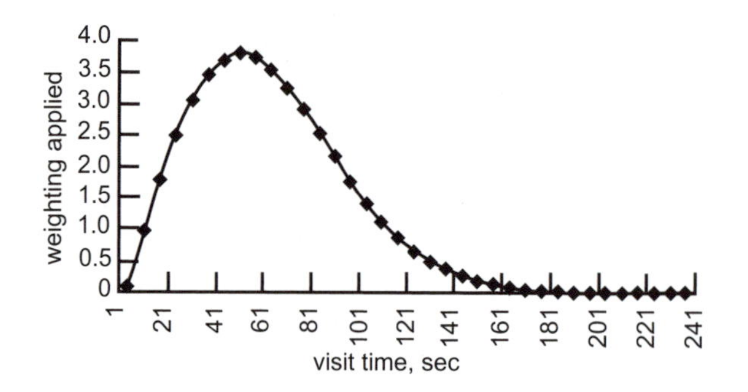

Fig 7.5 Keyword weightings, for a mean visit time of 60 sec.

If a person views a page for a very short time, the page is deemed to be of low relevance, and a lower weighting is given to all of the derived keywords. The same can be said for pages that are opened for a comparatively long period, which may occur if a window was opened but left in the background. Every time a change of Web page occurs, the weighting associated with a particular keyword is decremented. As a user moves over a series of Web pages, the weightings associated with each word will change, as well as the mean viewing time. When a keyword's weighting drops below zero, it is considered no longer to be in that user's current focus.

Figure 7.6 shows the weightings given to various keywords, during a particular user's browsing session. Given the scaling factor, K, as mentioned above, a word will stay in a current user's focus for approximately four Web page changes, but that is completely dependent on the time each page was visited, and whether that keyword occurs in more than one page.

keywords before	count
community	2
book	2
web	1
dynamic	1
framework	1
guidelines	1

keywords after	count
communities	5
journal	2
human/computer	2
preece	2
online	2
interdisciplinary	2
empathic	2

Fig 7.6 Changing focus during a browsing session.

A number of agent-based systems have been developed that provide information relevant to the user's context [25, 26]. Using the combined information across the whole community of browsers, Peruse provides feedback to the user on Web pages that are relevant, but have been seen by other members of the community. Also, Peruse will allow a user to find other people who are currently working with a similar focus to their own. Extensions to the current work would further extend the dynamic nature of Peruse, by adding features, such as instant messaging, in order to facilitate a more collaborative tool.

7.5 Delivering Relevance

Information which has been identified as timely and relevant needs to be delivered to users via an appropriate channel. Understanding the user's current activity is crucial in deciding whether a piece of information is relevant to the user, and hence whether the user should be informed about the information.

7.5.1 Explicit Searching

A classic case of not delivering relevance is the search engine. With the current generation of search engines:

- two users with different backgrounds and different interests will get the same results to an identical query;
- a user will (usually) get the same results to identical queries in the morning and the afternoon, even though they are working on completely different projects — moreover in the evening at home, the same results will be returned, even though the user is in 'leisure' mode;

- on revisiting a search engine query, the user will (again usually) be presented with identical results, even though the user has already investigated many of the recommendations.

As an example, consider multiple users searching the Internet for the word 'soap' using one of the many available search engines.

The Google search engine [27] returns over 3.6 million hits for the query 'soap'. But as a homonym, the word will have completely different meanings to the computer software engineer and the beautician. The computer software engineer, having spent the last hour browsing Web information about XML will no doubt be interested in the Google recommendations related to SOAP in the context of XML protocols, whereas the beautician will be interested in the version of soap that cleanses pores. Next generation search engines such as WiseNut [28] now provide rudimentary classifications of results from search queries, but still require the user to select the appropriate classification to drill down. This is a significant step forward, but still does not understand the user's context.

7.5.2 Personalised Search

With the user's dynamic profile as constructed above (section 7.4.2.2), it is now possible to retrieve more relevant results from search engines by including significant details of the user's dynamic profile in the search query.

Using the Google Web APIs service [29], which uses SOAP and WSDL standards, it is possible to access the Google search engine repository directly via their provided API.

In the prototype search interface developed, the user's search query is combined with (up to) the five most important terms from their dynamic profile, before being submitted to Google.

For example, in the 'soap' instance above, for a user who has been recently browsing a number of XML-related documents, the query to Google may take the form (in pseudo-code) of:

<must contain>soap
<may contain>XML
<may contain>software
<may contain>WSDL
<may contain>system
<may contain>W3C

On the other hand for the user who has been browsing a number of beauty-related sites, the query to Google may take the form of:

<must contain>soap
<may contain>skin
<may contain>bath
<may contain>beauty products
<may contain>shampoo
<may contain>glycerine

The number of matches returned by Google is unchanged by these additions, but the inclusion of terms from the user's dynamic profile will improve the search ranking of documents, giving higher weighting to those which include the dynamic profile terms.

Users are thus presented with search results which are more relevant to their current browsing activity. Figures 7.7 and 7.8 show the dynamic keywords extracted for a user, and the user's search results from our personalised search engine.

Your current keywords

keyword	count
skin	10
bath	8
beauty products	8
shower	7
glycerine	6
chemist	5
shopping trip	2
dry skin	2

Fig 7.7 Dynamic keywords.

7.6 Summary

Although the amount of information available to a user is growing, there is a potential to filter and deliver the appropriate level of information to a user based around an accurate, dynamic profile.

The development of tools such as Jasper, Peruse and dynamic profiling can proactively support the knowledge worker. By accurately representing the user's current context, it is possible to target relevant information to support the user's current activity.

Query results for soap

New Search Next >

Soap & Cosmetics on the Web
... Nuxe is a French natural cosmetics laboratory, which works to preserve **beauty**,
while it shows respect for the earth ... Conference Review in June, **Soap** & ...

soaps made by the **Soap** Box
... Moisturising and gentle to the **skin**. Makes a refreshing facial **soap** too.
... What you might want to know about our natural **beauty soap**... **Bath** and **beauty soap** ...

Handmade Soaps by Country Rose **Soap**. All Flowers and Herbs are ...
100% All Natural Handcrafted **Soap** & **Bath** Products. Enter Here.
Handcrafted all natural Herbal Soaps, **Bath** and **Beauty** Products. All ...

A Garden of **Soap** - for the Health and **Beauty** of your **Skin**
... nut oils (such as peanut oil, macadamia oil, etc) lanolin or soy oils because of the possibility of allergic reactions or **skin** irritation. At Garden of **Soap**, ...

GreenSense- Zenda's Homemade **Soap** and **Beauty** Recipes
... Zenda's Homemade **Soap** and **Beauty** Recipes, ... **Soap** Recipes - **Soap** Tips - **Bath** Oil, Salts and ... **Bath** Bombs. 1.4 cup baking soda 2 tbs ... this oil is absorbed into the **skin** ...

beauty & salon supplies
... offers several lines of natural, cruelty-free hair and **skin** care products, cosmetics, **bath**
... hair - provides **beauty** supplies and hair care products online. ...

Fig 7.8 Personalised search results for 'soap'.

References

1 InfoService: '*New model for the business portal*', Delphi (August 2002).

2 Lyman, P. and Varian, H.: '*How much information?*' — http://www.sims.berkeley.edu/research/projects/ how-much-info/

3 '*Glued to the Screen. An investigation into information addiction worldwide*', Reuters (1997).

4 '*Out of the Abyss: Surviving the Information Age*', Reuters (1998).

5 Davies, J. and Cochrane, R.: '*Knowledge discovery and delivery*', BT Engineering Journal (April 1998).

6 Davies, J., Krohn, U. and Weeks, R.: '*QuizRDF: search technology for the semantic web (PDF — 730 kb)*', WWW2002 workshop on RDF and Semantic Web Applications, 11th International WWW Conference WWW2002, Hawaii, USA (2002).

7 Davies, J.: '*Supporting virtual communities of practice*', in Roy R (Ed): '*Industrial knowledge management*', Springer-Verlag, London (2001).

8 Crossley, M., Davies, J., McGrath, A. and Rejman-Greene, M.: '*The Knowledge Garden*', BT Technol J, **17**(1), pp 76-84 (January 1999).

9 Wenger, E.: '*Communities of Practice: Learning, Meaning, and Identity*', Cambridge University Press (1998).

10 Jeffrey, P. and McGrath, A.: '*Sharing serendipity in the workplace*', in Proceedings of CVE 2000, San Francisco, CA, ACM Press, pp 173-179 (September 2000).

11 McGrath, A.: '*The forum*', Siggroup Bulletin, **19**(3), pp 21-25 (1998).

12 Cartwright, D.: '*DIY User Profiling*', Web Developer's Journal — http://www.webdevelopersjournal.com/articles/user_profiling_diy.html

13 Cartwright, D.: '*Customise Your Content with User Profiling*', Web Developer's Journal — http://www.webdevelopersjournal.com/articles/user_profiling.html

14 Bradley, K., Rafter, R. and Smyth, B.: '*Inferring relevance feedback from server logs — a case study in online recruitment*', Smart Media Institute, University College, Dublin — http://citeseer.nj.nec.com/382534.html

15 Greenspun, P.: '*Scalable systems for online communities*', MIT — http://philip.greenspun.com/panda/community

16 Avila, E. and Sherwin, G.: '*Profiles in courage*', — http://www.clickz.com/article.php/819341

17 Fernback, J. and Thompson, B.: '*Virtual communities: abort, retry, failure?*' — http://www.well.com/user/hlr/texts/VCcivil.html

18 Herzberg, A. and Ravid, Y.: '*Sharing bookmarks in communities*', IBM — http://www.ibm.com/

19 '*USA: BroadVision sees Java as the Future*', Computer Weekly, UK (June 2001).

20 Davies, N. J., Stewart, R. S. and Weeks, R.: '*Knowledge sharing agents over the World Wide Web*', BT Technol J, **16**(3), pp 104-109 (1998).

21 Brown, J. S. and Duguid, P.: '*Organisational learning and communities of practice*', Organisation Science, **2**(1) (1991).

22 Brown, J. S. and Duguid, P.: '*The Social Life of Information*', Harvard Business School Press, Massachusetts (2000).

23 Raybourn, E. M., Kings, N. and Davies, J.: '*Adding cultural signposts in adaptive community-based virtual environments*', in Purcell, P. and Stathis, K. (Eds): '*Intelligence and Interaction in Community-based Systems*', Special Issue of Interacting with Computers, **14**(6), Elsevier (2002).

24 Wolfram Research (2002) '*Eric Weisstein's World of Mathematics: Rayleigh Distribution*', — http://mathworld.wolfram.com/RayleighDistribution.html

25 Collis, J. et al: '*Living with agents*', BT Technol J, **18**(1), pp 66-67 (2000).

26 Underwood, G. M., Maglio, P. P. and Barrett, R.: '*User centered push for timely information delivery*', in Proceedings of the Seventh International World Wide Web Conference (WWW7), Brisbane, Australia — http://www.almaden.ibm.com/cs/wbi/papers/www7/user-centered-push.html

27 http://www.google.com/

28 http://www.wisenut.com/

29 http://www.google.com/apis/

8

DEVICE PERSONALISATION — WHERE CONTENT MEETS DEVICE

S Hoh, S Gillies and M R Gardner

8.1 Introduction

With such a diverse client base at present (which is expected to increase in the future), Internet service providers must consider how to create suitable content and a seamless service to the customer's preferred device. Present methods for determining client characteristics are ill defined and range from proprietary solutions to device-inferencing techniques. This is presently acceptable because, in part, most users' Internet experience is via the desktop computer, which has been the focus of most Web site and application design, and because the use of alternative devices such as personal digital assistants (PDA) and WAP-enabled mobile telephones has been limited. With predictions that mobile Internet access will exceed fixed line (desktop) during 2003, the demand for device-specific tailored content will increase and Web sites that can deliver such content will be the focus of much more Internet traffic.

The demands placed by this new generation of devices means that present solutions for determining client characteristics will be inadequate. Existing techniques, such as proprietary software (e.g. the AvantGo [1] browser) or inferring from information passed to the Web server (e.g. user-agent information included in the HTTP [2] packet header), are acceptable but not ideal. As the term implies, not all devices will support proprietary software and inferring characteristics from HTTP header information is problematic when devices with similar software have different characteristics or support for different media types. A better solution is to have the device characteristics passed to the server where information content can be tailored to the client's requirements. As more devices and services become available the industry will have to seriously consider how best to support such a widening range of devices. If not, there may be problems with interoperability/

compatibility between services, applications and devices. This may lead to a fragmented rather than unifying Internet experience for the majority of users.

8.2 Client Capabilities Recognition

To deliver the same information to different devices in a format that is suitable for its capabilities, an Internet application or Web site needs to be able to differentiate between the different device capabilities. The following are four possible methods for providing device recognition.

8.2.1 HTTP REQ Header Field

Web clients identify themselves when they send requests to Web servers. This identification is primarily for statistical purposes and the tracing of protocol violations, but can be used for personalisation. For example, early Netscape products generated user-agent strings such as: `Mozilla/4.04 (X11; I; SunOS 5.4 sun4m)` [3], which identified the requesting client device. However, these strings provided very little information and so are unsuited to precise content personalisation. A possible solution may be to extend the request header to accommodate additional client-capability information.

8.2.2 Proprietary Browser Header Field

Client properties can be relayed through the HTTP/1.0 and HTTP/1.1 user-agent headers in proprietary browsers [4]. This straightforward extension to HTTP is very simple but powerful enough to provide some client-specific information. The defined specification guarantees the upper compatibility with HTTP/1.0 and HTTP/1.1. Here, the client property information is added to a normal header using comment fields. However, the lack of standardisation between different proprietary browsers in the extension of the request header may prove a barrier for this technique to be used widely in content adaptation.

The following is an example of the type of additional information that can be added to HTTP requests:

Example: Internet-TV
User-agent: AVE-Front/2.0
(BrowserInfo
Screen=640 × 480 × 256;
InputMethod=REMOCON,KEYBOARD;
Page=512K; Product=XXXX/Internet-TV;
HTML-Level=3.2; Language=ja.JIS; Category=TV;
CPU=SH2; Storage=NO;)

8.2.3 Client Scripts

Scripting languages such as Jscript, JavaScript, VBScript and WMLScript can be used to provide device-specific information. There are also server applications that can interrogate devices. For example, Cyscape BrowserHawk [5] provides a scripting solution that interrogates the device and returns many of its characteristics including the connection speed.

8.2.4 Proprietary Diagnostic Application

Proprietary applications can be written (e.g. using ActiveX components) to report the browser device and connection characteristics.

8.3 The Client Profile

Device personalisation needs to consider the following three main elements:

- the device characteristics;
- the network connection or channel characteristics;
- the user's preferences.

8.3.1 Device Characteristics

Figure 8.1 shows a diverse range of devices that can connect to, and retrieve information from, the Internet. They all have common characteristics such as a screen, memory and microprocessor but these common characteristics are different for each device — screen size, memory size and processor power. The software support is also different and leads to issues including media (image, sound, etc), security (encryption standard, virus protection, etc), language (HTML [6], WML [7], JavaScript, etc), and protocol (WAP [8], HTTP2, etc) support. Clearly the requirements of each of these devices are different.

Fig 8.1 Devices that can retrieve information from the Internet.

8.3.2 Channel Characteristics

Figure 8.2 shows various possible Internet connection methods ranging from fixed line, wireless and mobile. The different types of connection, also known as the channel, for the most part perform the same function such as passing data to and from the connecting client. The channel characteristics differ, though, in a variety of ways, and the connection technologies vary within each method.

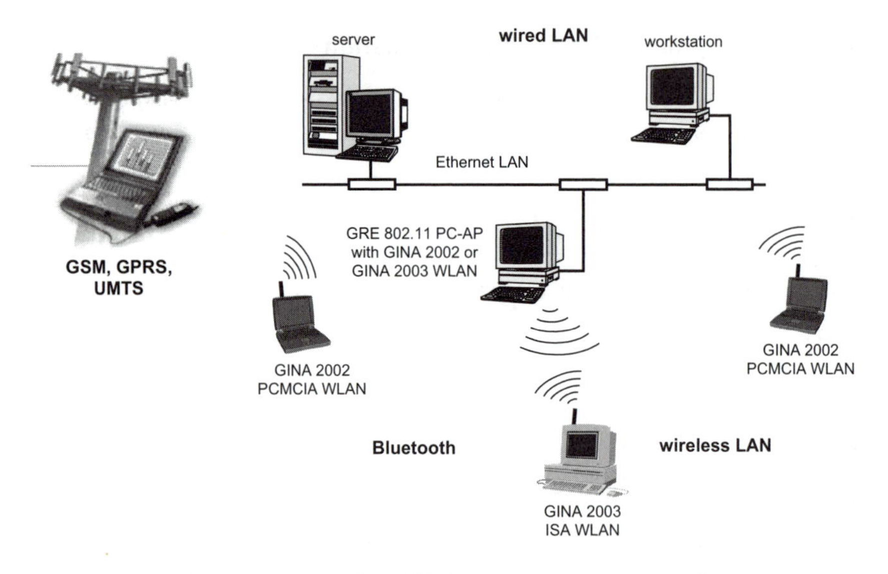

Fig 8.2 Range of possible Internet connection methods.

The most common types of network connection are summarised below:

- fixed line — standard narrowband modem, integrated services digital network (ISDN), digital subscriber line (DSL) and local area network (LAN);
- wireless LAN [9] — IEEE802.11b wireless LAN and Bluetooth personal area networks;
- mobile — GSM, GPRS and UMTS.

These technologies provide many different speed/bandwidth characteristics ranging from low bandwidth (9.6 kbit/s — 56 kbit/s), such as in cellular GSM and GPRS, and narrowband modems to middle bandwidth (128 kbit/s — 1.5 Mbit/s), such as ISDN, DSL, cellular UMTS and cable modems, to high bandwidth (10 Mbit/s — 100 Mbit/s), such as local Ethernet. The bandwidth on all these channels constantly varies and is dependent on the network and server traffic. There are also quality-of-service (QoS) issues to consider — for example, a LAN connection is generally more resilient than a public switched telephone network (PSTN) connection, which in turn is more resilient than a mobile connection.

8.3.3 User Preferences

All users are unique and their requirements can be based on the capabilities of the device (e.g. whether the particular device they are using at that time supports the form of media information they are trying to access), their own personal preference (e.g. no matter what bandwidth they have they only want text), and if they have a disability (e.g. a deaf person may want audio converted to text). Users may also have their own security requirements — the way that they identify themselves to the service provider may vary (such as SecureId tokens, username/password, digital certificates, etc).

8.4 The Standards — Binding the Profile

Several standards have been defined which address the interoperability problem. A few of these standards are described here.

8.4.1 WAP User Agent Profile (UAProf)

There is an extension of WAP 1.1 that enables a user agent profile (UAProf), also referred to as capability and preference information (CPI), to be transferred from a WAP client to a server [10]. This extension seeks to interoperate seamlessly with the emerging standards for CC/PP distribution over the Internet. The specification defines a set of components and attributes that WAP-enabled devices may convey within the CPI. This CPI may include, but is not limited to:

- hardware characteristics (screen size, colour capabilities, image capabilities, manufacturer);
- software characteristics (operating system vendor and version, support for MExE, list of audio and video encoders);
- application/user preferences (browser manufacturer and version, mark-up languages and versions supported, scripting languages supported);
- WAP characteristics (WML script libraries, WAP version, WML deck size);
- network characteristics, such as latency and reliability.

8.4.2 Composite Capability/Preferences Profile

The composite capability/preferences profile (CC/PP) standard describes a method for using the resource description framework (RDF) of the W3C, to create a general, yet extensible, structure for describing user preferences and device capabilities [11]. This information can be provided by the user to servers and content providers. The servers can use this information describing the user preferences to customise the service or content provided. The ability to reference profile information (RDF) via uniform resource identifiers (URIs) assists in minimising the number of network transactions required to adapt content to a device as well as reducing the header length. Further, the CC/PP framework fits well with the current and future protocols being developed at the W3C and the WAP Forum.

8.4.3 Resource Description Framework

The RDF is being used by the W3C as the foundation for the processing, exchange and reuse of metadata. Currently a W3C Recommendation, it aims to facilitate interoperability and modularity among disparate systems. The RDF allows statements to be made about resources, identified using a universal resource identifier. Each statement consists of three parts — the subject, the predicate and the object. The relationship between these statements is illustrated in Fig 8.3.

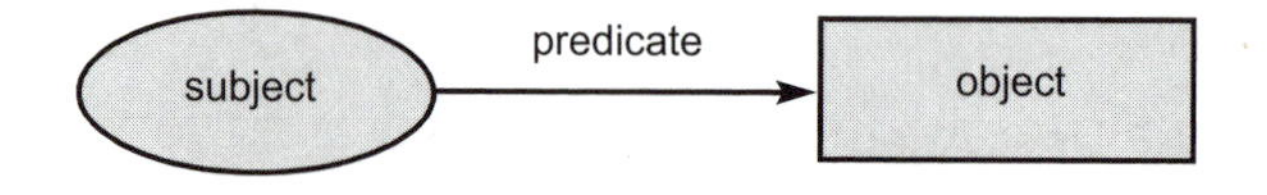

Fig 8.3 Relationship between RDF statement parts.

8.4.4 Current Implementations

There are several implementations of the CC/PP standard. Below is a list of a few of the current products.

- Commercial implementations:
  - — Nokia Activ Server UAProf Context Interface;
  - — Intel CC/PP SDK;
  - — Wokup! Server and Studio;
  - — Aligo M-1 Mobile Application Server;
  - — Brience Mobile Processing Server;
  - — Cysive Cymbio.
- Non-commercial implementations:
  - — DELI (HP Laboratories Open Source CC/PP Server API);
  - — DICE (University of Wales, Aberystwyth);
  - — UCP (IBM, Zurich);
  - — Jetspeed Capability API;

— XCries, CC/PP Servlet in XSmiles XML Browser;

— Panda and Skunk (Keio University).

Besides the currently available implementations, there is also an initiative within the Java Community Process (JCP) Program [12] which seeks to define a server-side API for the resolution support of client profiles based on the CC/PP standard [13]. This initiative aims to provide an extension to the J2EE platform to process delivery context information.

8.5 Content Negotiation — the Experiment

A project group at the BT Exact Asian Research Centre was mobilised to investigate and demonstrate how client device characteristics could be captured and used for content negotiation. The devices chosen for the demonstrator were a standard PC, a PDA and a WAP mobile. A demonstrator was built, consisting of the five core components[1] listed below:

- discovery module;
- client proxy;
- profile registration agent (server);
- profile servlet;
- demonstration application.

8.5.1 Hosting the Profiles

Initially, a downloadable discovery module interrogates the client to elicit its capabilities. Immediately after obtaining the profile, it is sent to the profile registration agent which, in turn, registers the profile in a server database. In this demonstrator, the profiles are hosted on a server especially developed for that purpose within BT.

8.5.2 The Profile Referencing Mechanism

After storing an entry, a reference URI is returned to the client to be used. This URI then forms part of the CC/PP profile mechanism within the request header. In order to alter the outgoing request header, a client-side proxy is used. This traps all requests going out from the browser, modifying them to provide client-side CC/PP compliance.

Table 8.1 shows an example REQ header before and after proxy modification.

[1] Some of these components do not apply to WAP devices as existing UAProf support on these devices already satisfies some of these requirements.

Table 8.1 REQ header before and after proxy modification.

Before:
```
GET /ezassist/ HTTP/1.1
Accept: *.*
Accept-Language: en-gb
```

After:
```
GET /ezassist/ HTTP/1.1
Accept: *.*
Accept-Language: en-gb
x-wap-profile:
"http://localhost:8080/ProfileServlet/testing/test1234/Default","1-
h7ag3PSANDtaDBW4z5611w=="
x-wap-profile-diff:1; <?xml
version="1.0"?>
<rdf:RDF xml:lang="en"
xmlns:rdf="http://www.w3.org/1999/02/22-
rdf-syntax-ns#"
xmlns:rdfs="http://www.w3.org/2000/01/rdf-
schema#"
xmlns:prf="http://www.wapforum.org/profiles/
UAPROF/ccppschema-20010430#">
        <rdf:Description
        rdf:ID="MyDeviceProfile">
        <prf:datagramreceived>
        81837
        </prf:datagramreceived>
.....
.....
</rdf:RDF>
```

8.5.3 Profiles Hosted — So What?

Now, when a user makes an HTTP request to a service provider, it will include a URI pointing to a detailed profile of the device. The service provider can process each REQ header to extract this URI and the device profile. The profile and its difference are then resolved to obtain an up-to-date client profile which describes its capabilities.

Figure 8.4 illustrates the process flow of identifying, registering, and usage of the profile.

8.5.4 Negotiating the Content Source

Once the profile has been obtained, it can then provide the user with information that is tailored to the client device and preferences. There are several techniques

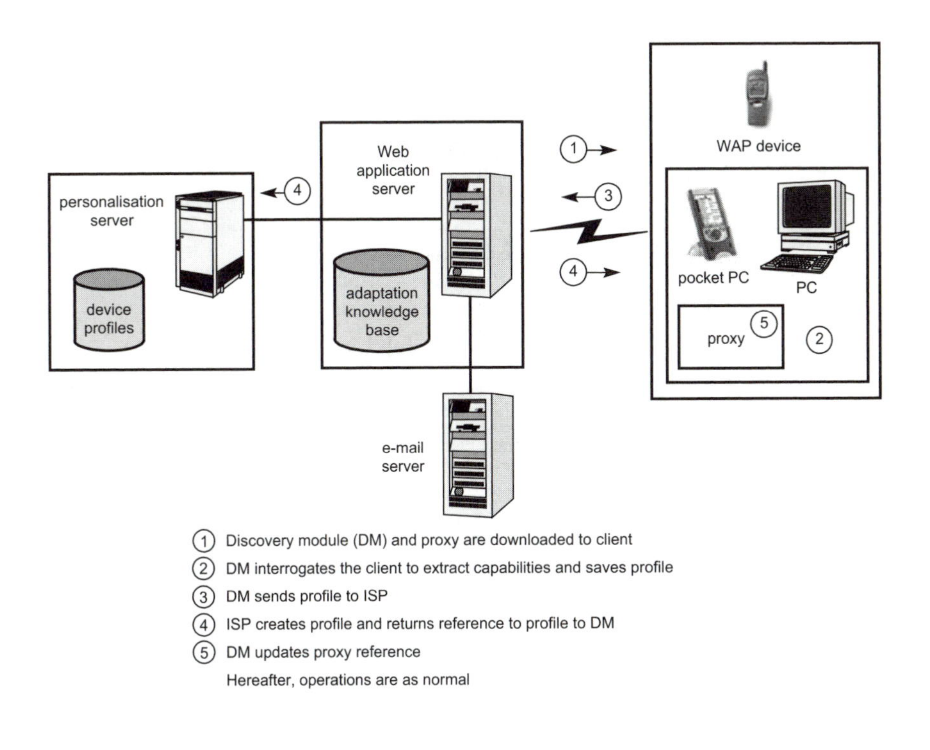

Fig 8.4 Process flow of identifying, registering and usage of the profile

which could be used for content adaptation [14]. These techniques could be classified into five categories:

- purpose classification;
- information abstraction;
- modality transformation;
- data transcoding;
- data prioritisation.

Table 8.2 lists the content adaptation techniques used in each of these categories. For the purpose of this experiment, a Web-based application has been created to demonstrate how Web content could be easily adapted for each of the connecting devices using their respective client profiles. The demonstrator focuses only on adaptation of text and images. Nevertheless, similar adaptation techniques could be applied across all media types.

Table 8.2 Content adaptation techniques.

	Video	Image	Audio	Text
Purpose classification	Removal Substitute	Removal Substitute	Removal Substitute	Removal Substitute
Information abstraction	Video highlight Video framerate reduction Video resolution reduction Keyframe extraction	Image dimension reduction Data size reduction (increased compression rate)	Audio highlight Audio sub-sampling Stereo-to-mono conversion	Text summarisation Outlining Font size reduction Text whitespace removal
Modality transform	Video-to-image Video-to-audio Video-to-text Removal	Image-to-text Removal	Audio-to-text Removal	Text-to-audio Removal Table-to-list Table-to-plain text Language translation
Data transcoding	Format conversion Colour depth reduction	Format conversion Colour depth reduction Colour to gray-scale	Format conversion	Format conversion
Data prioritisation	Layered coding Frame priortisation Audio prior to video	Multi-resolution image compression	Audio prior to video	Text prior to image/audio/ video

8.5.5 The Result

Using the open-source Web publishing framework Cocoon [15], it was possible to segment the styling information from the data. The styling information is then customised, based on the respective client profiles, and is structured to form the necessary output mark-up using the extensible stylesheet language (XSL). The client profiles are made visible to stylesheets as parameters. Hence, within each stylesheet, the client profile attributes could be queried and used appropriately to style a content page to suit the corresponding device.

Figure 8.5 illustrates how the first page of the demonstrator is served to each individual client device.

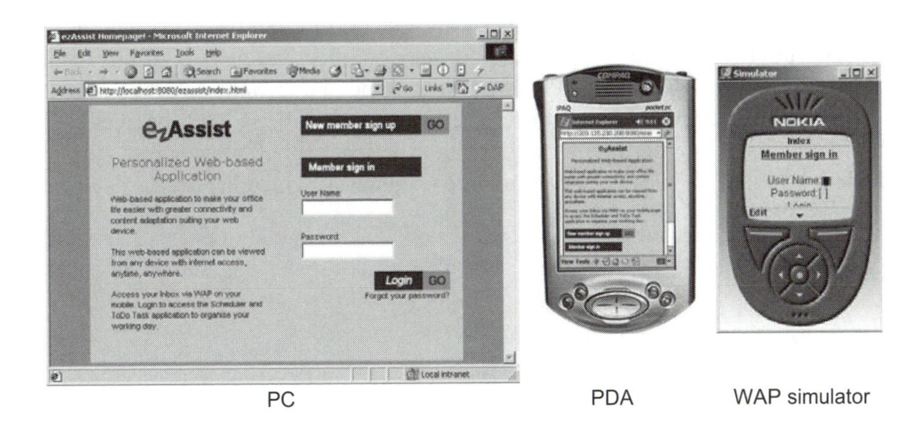

Fig 8.5 Illustration of how the first page of the demonstrator is sent to each individual client device.

8.6 Summary

Determining the characteristics of a client connecting to a Web application is manageable with the present technologies. Device attributes are obtainable through a variety of methods but currently the most successful is from information contained in the HTTP packet header, whether this is proprietary or by inference. It should be noted that inferring characteristics may not always be successful as the same information may, especially in future clients, be similar in different devices with correspondingly different characteristics. The use of scripting technologies works well with devices that support the technology but users can disable this feature — so this is generally not a dependable technique. The channel profile of the connecting client is very difficult to determine from an initial client connection, but it is possible to determine speed after a designated amount of traffic has passed between client and server. Due to the variable nature of the traffic on the Internet a balance has to be made between a default speed and constant recalculation of the channel speed — however, this may increase the server load especially if there are many connecting clients.

Usually, the user profile is not known at the initial connection unless the service or application has prior knowledge of the user. This is generally only possible if the user connects using a registration interface or a cookie that has been previously saved on the client device. The most effective method for determining the client profile will be the implementation of the joint W3C, IETF and WAP proposed CC/PP standard by the Internet industry. There are many issues involved but the crucial one is the adoption by manufacturers of this standard. This is difficult when many manufacturers are producing proprietary solutions for their own products in the hope that they can build up a customer base, and so be first to market, and hence persuade Internet application providers to support their product. If a universally agreed solution within the Internet industry is not found there may be problems with interoperability/compatibility between services, applications and devices. The ultimate aim is to provide a unifying Internet experience for all users across multiple (device) channels — be it PC, PDA, mobile telephone or any other access device.

References

1 AvantGo — http://www.avantgo.com/

2 HTTP — http://www.w3.org/Protocols/

3 Zawinski, J.: '*User-agent strings*', — http://www.mozilla.org/build/user-agent-strings.html

4 Kamada, T. and Miyazaki, T.: '*Client-specific Web services by using user agent attributes*', W3C — http://www.w3.org/TR/NOTE-agent-attributes-971230.html

5 BrowserHawk — http://www.cyscape.com/

6 HTML — http://www.w3.org/MarkUp/

7 WML — http://www.wapforum.org/

8 WAP — http://www.wapforum.org/

9 Wireless LAN alliance — http://www.wlana.org/

10 Open Mobile Alliance: '*WAG UAProf*', Version 20-Oct-2001 — http://www.openmobilealliance.org/wapdocs/wap-248-uaprof-20011020-a.pdf

11 Klyne, G., Reynolds, F., Woodrow, C. and Ohto, H.: '*Composite capability/preference profiles (CC/PP): structure and vocabularies*', W3C Working Draft (March 2001) — http://www.w3.org/TR/2001/WD-CCPP-struct-vocab-20010315/

12 Java Community Process — http://www.jcp.org/

13 JSR-188 CC/PP Processing — http://www.jcp.org/jsr/detail/188.jsp

14 Wei-Ying, M., Ilja, B., Grace, C., Kuchinsky, A. and Zhang, H.: '*A framework for adaptive content delivery in heterogeneous network environments*', SPIE Multimedia Computing and Networking 2000, San Jose, pp 86—100 (January 2000).

15 Cocoon — http://www.apache.com/

Bibliography

Benkiran, A. and Ajhoun, R.: '*Towards an adaptative and cooperative tele-learning*', Proceedings of the International Conference on Engineering Education, Session 7B1, pp 22-27 (2001).

Bharadvaj, H., Joshi, A. and Auephanwiriyakul, S.: '*An active transcoding proxy to support mobile Web access*', in 17th IEEE Symposium on Reliable Distributed Systems (October 1998).

Butler, M. H.: '*Current technologies for device independence*', HP Labs Technical Report HPL-2001-83 — http://www.hpl.hp.com/techreports/2001/HPL-2001-83.html

Byun, H. E. and Cheverst, K.: '*Exploiting user models and context-awareness to support personal daily activities*', Workshop in UM2001 on User Modelling for Context-Aware Applications, Sonthofen, Germany (2001).

Chakrabarti, S., Mukul, M. J. and Vivek, B. T.: '*Enhanced topic distillation using text, markup tags, and hyperlinks*', ACM Special Interest Group on Information Retrieval, New Orleans (2001).

Cyrus, S. and Farnoush, B. K.: '*Efficient and anonymous Web usage mining for Web personalisation*', in INFORMS Journal on Computing, Special Issue on Data Mining (2002) — http://dimlab.usc.edu/Publication.html

Fox, A., Gribble, S. D., Brewer, E. A. and Amir, E.: '*Adapting to network and client variability via on-demand dynamic distillation*', Proc Seventh Intl Conf on Arch Support for Prog Lang and Oper Sys (ASPLOS-VII), Cambridge MA (October 1996).

Fox, A., Goldberg, I., Gribble, S. D., Lee, D. C., Polito, A. and Brewer, E. A.: '*Experience with Top Gun Wingman: a proxy-based graphical Web browser for the USR Palmpilot*', Proc IFIP International Conference on Distributed Systems Platforms and Open Distributed Processing (Middleware '98), Lake District, UK (September 1998).

Joshi, A.: '*On proxy agents, mobility, and Web access*', ACM/Baltzer Journal of Mobile Networks and Applications, **5**(4) (2000).

Katz, R. H.: '*Adaption and mobility in wireless information systems*', IEEE Personal Communications, **1**(1), pp 6-17 (1994).

Kirda, E., Kerer, C. and Matzka, G.: '*Using XML/XSL to build adaptable database interfaces for Web site content management*', XML in Software Engineering Workshop (XSE 2001), 23rd International Conference on Software Engineering, Toronto, Ontario, Canada (2001).

Korolev, V. and Joshi, A.: '*An end-to-end approach to wireless Web access*', in Proc of the 21st IEEE International Conference, Distributed Computing Systems Workshops, IEEE Computer Society, pp 473-478 (April 2001).

Lakshmi, V., Ah-Hwee, T. and Chew-Lim, T.: '*Web structure analysis for information mining*', Proc of the ICDAR'01 Workshop on Web Document Analysis, Seattle (2001).

Ramakrishnan, N.: '*PIPE: Web personalisation by partial evaluation*', IEEE Internet Computing, **4**(6), pp 21-31 (November-December 2000).

Seshan, S., Stemm, M. and Katz, R. H.: '*SPAND: shared passive network performance discovery*', in Proc 1st Usenix Symposium on Internet Technologies and Systems (USITS '97) Monterey, CA (December 1997).

Smyth, B. and Cotter, P.: '*Content personalisation for WAP-enabled devices*', in Proc of Machine Learning in the New Information Age Workshop (MLnet / ECML2000), pp 65-74, Barcelona, Spain (2000).

SyncML Initiative, Ltd: '*SyncML device information DTD, v1.1*', — http://www.syncml.org/docs/syncml_devinf_v11_20020215.pdf

9

MULTILINGUAL INFORMATION TECHNOLOGY

S Appleby

9.1 Introduction

The ubiquity of the World Wide Web (WWW) has given us the means to access information from all around the world — if only we could understand the language in which it is written!

Currently, non-English-speaking World Wide Web users outnumber English-speaking users, with the number of non-English users growing at a far greater rate (see Fig 9.1). Particularly fast growth is seen in China, Brazil and Korea. Providers of Internet-based services will have to become multilingual and multicultural if they are to capitalise on this huge increase in potential market.

Internet portal providers were among the first to recognise the value that can be added to their services by providing facilities, such as automatic translation, to help users cope with information in other languages.

For example, AltaVista [1] allows automatic translation of Web pages using the Systran translation system [2, 3], and Yahoo [4] allows translation using the Logos translation system [5].

Making an application or service work in different languages may involve much more than just translating some of the text. In many cases, applications would benefit from being ported to other locales, but the cost is prohibitive because the system was designed originally for monolingual use, and it is now too expensive to re-engineer it. For example, mistaken choice of character encoding scheme at an early stage could make a significant difference to the cost of porting to other languages.

The aim of this chapter is to present a very brief overview of some of the issues in producing applications that work in different languages. Firstly, some of the general issues in application design are discussed, and then the chapter focuses more specifically on two areas — display of text and automatic translation.

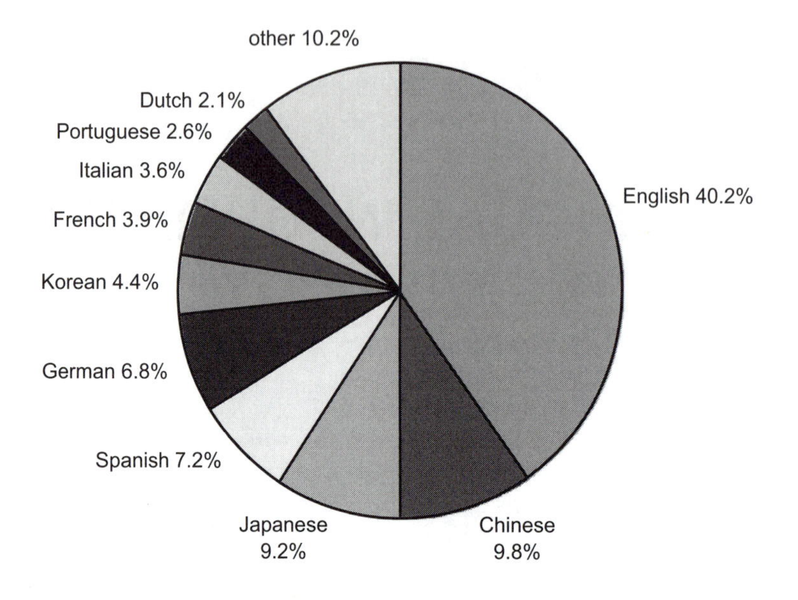

Fig 9.1 Number of World Wide Web users by language (March2002). [Source: Global Reach — International Online Marketing [6]]

9.2 Localisation and Internationalisation

Localisation is the process of taking a product and converting it for sale and use in another locale — where the term locale refers not only to regions that are distinguished by speaking different languages, but where anything that can influence the product or its marketing or sales process may be different. Localisation includes (but is certainly not limited to) the translation of all text associated with a product (such as on-line help, user manuals, dialogue boxes, etc), as well as changing currency, date notation, perhaps even porting the application to a different version of an operating system. Culture can also play a significant role in localisation, e.g. in some regions, the 'cute' icons, which first started to be used on Macintoshes, are not well received [7].

Internationalisation is the term used to refer to the process of designing an application or service in such a way as to prepare it for efficient localisation.

Computer operating system producers were among the first to realise the importance of internationalising their products. They realised that they had to consider multiple markets at every stage of designing their systems. Otherwise, porting to other locales may be prohibitively expensive or delay product introduction to such an extent as to reduce market share.

Now, facilities to help with internationalisation are available at all levels of application development. Computer programming languages such as Java have features which make internationalisation much easier. With the advent of HTML 4 and its support by the major Web browsers, WWW pages can now incorporate multiple languages, even on the same page [8, 9].

9.2.1 Internationalisation and System Design

Systems must be designed and built with internationalisation in mind throughout [10]. It is very easy for a development team to incorporate language-specific features in a system without realising that this is happening. Their minds are usually focused more on the application logic, and the design that is most easily ported to another locale is not necessarily the most elegant to a system designer.

For software systems, globalisation influences several aspects of the product, such as:

- text that will appear in the user interface;
- graphical user interface design;
- help files, manuals and technical support;
- locale-specific data (e.g. product databases);
- application design (in some cases).

Clearly text that appears on the user interface will need to be translated. To facilitate this, it is now well known that software writers must not embed text in their application, but must instead store text as external resources that can be translated without needing to recompile the system. Ideally, translators would have access to a running version of the system so that they can be certain of the context and they can see how their translated text appears on the screen.

The user interface designer will need to bear in mind that English text is quite compact compared with other languages. The designer should typically allow for a 30% to 50% expansion for text translated into other languages that use the Roman script. For languages which use scripts other than Roman, there are further difficulties. Languages whose script is based on Chinese Han characters (Chinese, Japanese and Korean (CJK)) traditionally run top to bottom and right to left. Fortunately for system designers it is now common to see such scripts run left to right and top to bottom as for Roman script. However, the typesetting and font design requirements of CJK scripts are completely different from those of the Roman scripts [11].

Translation of help files and manuals is a more or less standard text-translation process. However, it should be borne in mind that the English documentation (assuming English is the original language) will invariably be read by users for whom English is not their first language. This means that the style should be as free from ambiguities as possible — even to the point of being more repetitive than would be preferred by an English speaker.

Some companies (such as Caterpillar, Perkins, and many others) employ so-called controlled languages for producing manuals [12, 13]. These are formally prescribed subsets of (usually) English which are less ambiguous and more consistent than normal language. The reduced ambiguity not only makes the documents clearer to non-native speakers, but also reduces the time and cost of the translation process.

The cost of translating manuals and help files can also be reduced by using a translation memory. This will remember previous translations and can either suggest them to the human translator, or can insert them automatically. When producing a new version of a set of help files, using a translation memory will avoid the need to re-translate those parts that remain unchanged. If a translation memory is used in conjunction with a controlled language, there is a far greater chance of repetition, and therefore of the translation memory being able to translate more of the text automatically. Translation memories also encourage a more consistent use of terminology, especially when the memory is shared among several translators working on the same project.

There are other utilities that can help with software localisation, these will help to manage resource files, design menus and dialogue boxes, etc, and extract text for translation.

An example where language may influence the design of the application, is where a search engine is used. Search engines are used to retrieve data, normally text, based on the keyword searched. Most search engines are monolingual. Porting such a monolingual search engine to another locale will, as a minimum, involve changing the morphological rules used to derive the base forms of words from their inflected forms.

More sophisticated search engines may use a knowledge-base or thesaurus for keyword expansion and disambiguation, or may use some grammatical analysis of text. It may be that some retrieval techniques that are found to work in one language will not work in another.

Languages with no clear word boundaries will present particular problems to search engines. For example, Chinese, Japanese and Korean have no markers to indicate word boundaries (an equal space is inserted between each character).

A more complete internationalisation of a search engine would involve cross-linguistic retrieval — that is where a query made in one language could retrieve documents stored in another.

One technique typically used in internationalising applications is to use a level of indirection to refer to locale-sensitive information. The most common case of this is when resource files are used. A resource file would contain all the text, dialogue box parameters, etc, that are specific to a particular locale. The application is localised by installing the appropriate resource file.

An analogous indirection technique may be used in real time. For example, multinationals often have a requirement for multilingual and multi-locale customer relationship management (CRM) solutions.

The operational CRM application will need to have access to information that is sensitive to the customer context, including locale-specific information. A 'language context object' could be employed to select the locale-specific information for a given customer.

9.3 Text Display

The vast majority of user interfaces have to be able to display text. In an international context, this is not always as easy as English speakers assume. This section will give an overview of the difficulties of displaying text in languages that use complex scripts.

9.3.1 Characters, Glyphs, Scripts and Fonts

A glyph is the visual representation of a character. Informally, we often misuse the term 'character' by using it to refer to a glyph. A character is an abstract notion. A character cannot be displayed, only glyphs can be displayed. A character encoding scheme is a function which maps from integers to characters (it is possible for two different numbers to be assigned to the same character, but not for a number to be assigned to two different characters).

A font provides a mapping between character codes and glyphs. We typically specify a font family, a typeface and character size, etc, to provide the remaining information necessary to map character codes on to glyphs.

A text rasteriser will take a font and a sequence of character codes and produce a graphic illustrating the required glyphs.

The relationship between characters and glyphs is sometimes not a straightforward one. We would normally assume that a given letter rendered in say, italic and bold, is the same character represented by two different glyphs. It is also normal to assume that upper and lower case versions of the same letter are different characters, although the requirement for this is probably more of a practical nature, since we do not want to leave the decision of case to the rasteriser.

Other languages present more of a problem though. In Arabic each letter typically has a number of written forms which depend on the location of the letter within a word. Letters have quite different representations, i.e. quite different glyphs, depending on whether they are at the beginning, middle or end of a word, or are written in isolation. Should we consider each of these forms to be different characters, in the same way that we might consider upper and lower case letters to be different, or should we assign all the different forms of the same letter a different character code?

If we assign a single character code to every form of a given letter, then it would be left to the text rasteriser to decide which glyph to display in a particular context.

Somehow, the rasteriser would have to have access to language-specific rules telling it how to make this decision.

In practice, Arabic encoding schemes include special codes for so-called 'presentation' forms. These are codes assigned just for the different variants of a given character. This means that the character codes begin to look more like glyph codes than character codes.

In Arabic, we can get away with this solution, because the number of presentation forms is small and reasonably well-defined. In other languages, such as those written in a Sanskrit-based script (such as Hindi, Gujarati, Bengali, etc), or Urdu, the problem is much more difficult to solve. Such scripts are complicated by the fact that the way that a letter is written, and the position in which it is written, are highly dependent on context. Figure 9.2 shows an example of how a sequence of isolated letters should be rendered by the rasteriser in Hindi.

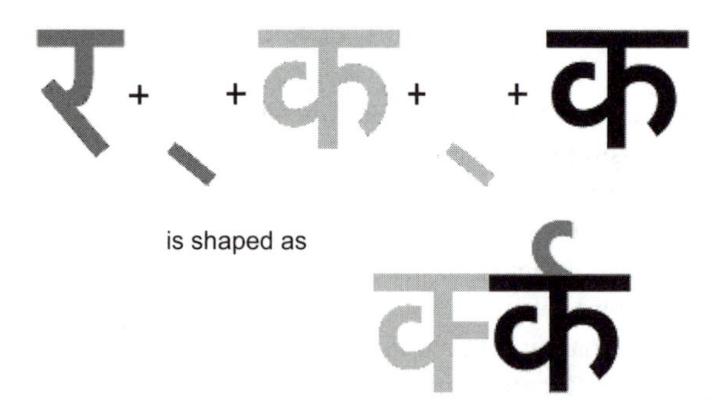

Fig 9.2 An example of a conjunct in Hindi. A text file would contain the character codes for the five glyphs on the top line, and the text rasteriser would be expected to display this as the conjunct shown on the bottom line.

It is very difficult to assign a character code to each possible presentation form since there is no fixed set of these. The number of presentation forms used will depend on the designer of the font and the text processing package being used. Older or more formal languages tend to favour more complex conjuncts, whereas more modern texts tend to use conjuncts less.

Urdu presents an even bigger challenge. Urdu is traditionally written in a very ornate script called Nastiliq, which is derived from Arabic. If we take the Unicode character encoding scheme as an example, we find that Urdu shares the Arabic code pages. This means that there are a few hundred codepoints assigned for Arabic and Urdu. A high-quality Urdu font will define several thousand (perhaps up to 20 000) glyphs. This is, again, because the presentation form of a letter will depend on its context, and several letters together may be represented by special glyphs.

So, how are these problems solved in practice? Manufacturers of Indic fonts and text-processing software invent their own character-encoding schemes that, in fact, look more like glyph encoding schemes. They assume a one-to-one mapping between character codes and glyphs (which removes any responsibility from the rasteriser for deciding which glyphs to display).

The result of this is that currently available fonts do not use standardised encoding schemes for these languages. This makes it very difficult to exchange documents electronically, since you have to ensure that everyone has access to the same text-processing software. It also makes display of Web pages difficult, since the display of text usually relies on fonts installed on the client. With no standardisation, it is very unlikely that the client will have installed exactly the same font, from the same manufacturer.

Sometimes font manufacturers claim to adhere to the Unicode standard. However, they make heavy use of the private code area to encode the various presentation forms. This means that different manufacturers can claim to adhere to the Unicode standard and yet produce fonts with incompatible encoding schemes.

As a solution to these kinds of problem, Microsoft has created the OpenType font standard. Essentially, OpenType introduces extra tables, in addition to those in the more familiar TrueType font standard, which allow the font manufacturer to define the mapping between sequences of character codes and sequences of glyphs. The text rasteriser needs to understand these rules, but does not need to have any language-specific information coded into it.

It is clear that, especially in a multilingual context, differences between the fonts used by the author of a document and those used by the reader could cause a document to display incorrectly, to the point where text is not displayed at all. Typically when designing Web pages, only a small number of common fonts are used so that you can have a reasonable confidence that your page will display on other screens roughly the way that it appears on yours. Ideally, you would like to be certain that everyone who views your Web pages has the same fonts installed that you have on your machine.

Two competing technologies are now helping to make this possible. Bitstream [14] has produced a system called TrueDoc, which is supported by Netscape (among others), and Microsoft [15] and Adobe [16] have produced OpenType, which is now probably the dominant technology. Both of these systems allow the author of a page to bundle their fonts along with the page in a highly compressed form. When someone accesses the page, the font rendering information will be downloaded with the page and so characters will be displayed in the way the author intended. The World Wide Web consortium [7] is considering adoption of a single standard to allow portable fonts to be used. For the time being, Web page designers will have to resort to inelegant solutions to make sure that both Netscape and Internet Explorer will show pages properly. For simpler scripts, font sets are beginning to appear which contain a reasonably comprehensive implementation of the characters in the Unicode set. For example, Bitstream's Cyberbit font implements some 39 000 characters from the Unicode range [14].

9.3.2 Input Method Editors

Text entry poses a different set of problems to text display. For languages with relatively small character sets, we are used to having a keyboard with all the relevant characters displayed. In other languages, most notably those with Chinese-derived writing systems, this is not practical.

For these languages, a software utility is used which intercepts keyboard input to allow the selection of the appropriate characters. Such a utility is called an 'input method editor' (IME).

Typically, a Chinese speaker may enter a number of roman characters (in a system called Pinyin) which together make up a phonetic representation of the required Chinese character. As there are normally a number of Chinese characters for a given Pinyin transcription, the IME will present the user with a list of options. The user will then select the appropriate character and the IME will insert it into the text.

Although phonetic-based entry methods are the most common, there are other methods (such as stroke-counts) for identifying characters. IMEs normally include a range of methods.

9.4 Machine Translation

Clearly there is a huge financial incentive to be able to translate from one language to another fully automatically. In this section, we discuss some of the challenges involved in building an automatic translation system.

At its highest level, the art of translation is a very subtle and creative one. Machine translation techniques do not attempt to reproduce this quality, instead they aim for intelligibility, factual accuracy, and if possible, an appropriate style or register (level of formality) [17].

Research on machine translation started in earnest in the 1940s with a memorandum from Warren Weaver. Initially translation was conceived as a problem that could use the code-breaking techniques developed during the war. The idea was that you considered a message that was written in a language that you could not speak as having originally been written in your own language, it is just that someone encrypted it first. To translate you 'simply' have to decrypt the encrypted message.

It was quickly realised that this technique does not work very well. The main reason is that human languages make very poor coding schemes.

This is primarily because most words have multiple meanings, but also because the structure of sentences is not uniquely recoverable from a given linear sequence of words.

9.4.1 Difficulties with Translation

9.4.1.1 Ambiguity

There are various kinds of ambiguity. The most obvious is word sense ambiguity. Take any common word, look it up in a good dictionary and you will find several meanings for it. The more common and apparently insignificant the word, the more likely it is to have a long entry in the dictionary. Words like 'of' or 'be' will usually have many senses listed. Word sense ambiguity causes two problems to the designer of a machine-translation system:

- how to represent the fact that the same word may have several meanings;
- how to select the appropriate meaning.

Simply listing the possible senses of a word is one way to represent the different meanings of a word. However, it is not clear how many entries a word should have in order to characterise all its possible senses. For example, if we are considering the dictionary entry for a word such as 'bank', we might decide that there are the following senses:

- a financial institution;
- the side of a river;
- to place one's financial business with a financial institution;
- to rotate (as of an aeroplane);
- various colloquial phrases ('to bank on something').

Remembering that our aim is to translate, we find that when we want to translate the word 'bank' into French, it would be useful to know whether it was a high-street bank or a merchant bank (caisse d'épargne, banque). In this case, the first sense of bank could be considered to be further subdivided. So for translation, whether a word appears to be ambiguous or not in the source language could depend upon the granularity of the distinctions that are made in the target language.

Word sense ambiguity can become more subtle if we allow the distinction between, say, the financial institution and the building containing the financial institution. For example, if I say 'my bank refused to extend my overdraft', I would be referring to the institution (or perhaps more correctly, to someone speaking on behalf of the institution). If I said, 'A bolt of lightning hit the bank', I would be referring to the building.

Another kind of ambiguity is 'structural ambiguity'. Structural ambiguity is where, even if the precise meaning of each word in a phrase is known, it is not possible to tell what the structure of the sentence is, and so you cannot interpret it properly. For example, these two sentences have two different syntactic structures:

- I went to the park with swings;
- I went to the park with my brother.

In the first case, 'the park' is associated with 'swings', in the second case either 'I' or 'went' is associated with 'my brother'.

When languages are very similar in structure, say English and French, it may be possible to translate a text even if its syntactic structure cannot be resolved. In general though, syntactic structure will be important for choosing the right translation.

9.4.1.2 Semantic Spaces

Another difficulty with translation is that different languages associate words with different concepts [18], i.e. their partitioning of reality may be different. This means that considerable paraphrasing will need to be carried out to be able to produce a text which conveys roughly the same meaning as the original. Such paraphrasing could require extensive background knowledge [19].

Related to the problem of semantic spaces is that of missing vocabulary. Consider translating the phrase 'how quickly?' from English into French. French has no equivalent for the word 'how' in this context. To find a translation, it is necessary to paraphrase 'how quickly?' as 'à quelle vitesse?' (which is literally 'at what speed?'). To be able to carry out such paraphrasing in a general way, the translation system will need to know that there is a relationship between the concepts associated with 'quick(ly)' and 'speed'. This could mean that the translation system would have to store numerous relations between concepts in order to have sufficient knowledge to paraphrase where directly corresponding words are not available in a particular language.

9.4.1.3 Other Aspects of Translation

Suppose we wish to compare two texts in two different languages to say whether they represent accurate translations of one another. We might compare them in terms of their truth conditions. That is, in a purely logical sense, are the statements made in the texts true and false in exactly the same situations? For example, the statement 'Romeo loves Juliet' is true in exactly those situations in which 'Juliet is loved by Romeo'. So a translation which switches from active to passive mood would be accurate in a truth conditional sense.

Language is a communication medium, and as such to decide what a good translation is, we need to go beyond truth. The theme of 'Romeo loves Juliet' is Romeo, since this is a statement about Romeo. However, the theme of 'Juliet is loved by Romeo' is Juliet, since this is a statement about Juliet. The communicative value of these two statements is therefore not equivalent [20]. A translation which is sensitive to these communicative aspects of language will clearly be better than one that is not.

Normally, theme is marked in English using word order. In other languages, such as Japanese, theme is marked using a particle (a small word that serves as a marker but has no other meaning of its own). To translate with any kind of accuracy between English and Japanese, it will be necessary to identify the communicative function of a phrase as well as its logical content.

There are still more subtle aspects to communication, such as choosing the right register and adjusting the rhythm of a particular piece of language. When adding subtitles to a film for example, it is necessary to choose the translations in such a way as to try to keep the flow of the subtitles in step with the images.

9.4.2 Approaches to Machine Translation

9.4.2.1 Direct Approach

The most obvious way to translate a text might be to use a bilingual dictionary to look up words in the source language and replace them with words in the target language. Of course, such a technique will produce very poor translations. Firstly, such a system has no knowledge of the target language grammar, and so the target text will be ungrammatical. Secondly, it has no knowledge of the source language grammar or context, and so word selection will be very poor.

A simple approach such as this can be modified to include rules to help disambiguate the source text and re-order target words. It can also be extended by providing translations of phrases for common cases where single word translations fail.

Such was the design of one of the earliest successful machine translation systems, Systran [2], which is now best known as the translation engine behind the AltaVista Babelfish.

This approach is essentially a huge dictionary with guidelines that say how each word should be used. To achieve a reasonable coverage and translation accuracy, each entry in the dictionary needs to include a set of instructions saying how the word should be translated for each possible context. Such a task would be endless, so, in practice, economic realities constrain such systems to have poor accuracy.

9.4.2.2 Transfer Approach

The direct approach is entirely lexicalised, i.e. there is no explicit grammar. A grammar would be useful in representing many generalisations though, and would reduce the number of rules that are required. In the so-called 'transfer' approach, a grammatical analysis of the source text is carried out, and then transfer rules are applied to transform the source analysis into a suitable structure for generating target text. The transfer approach is shown in Fig 9.3.

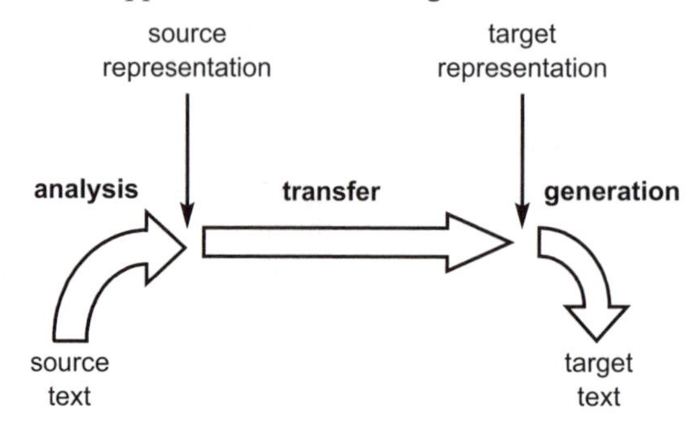

Fig 9.3 The translation process in a transfer type translation system.

In practice, the generation stage is often very simple, most of the target grammar being encoded as transformations of the source analysis (e.g. IBM's LMT system [13, 21]).

Most current higher-performance translation systems now use, or are aspiring to use, some version of the transfer approach. The main disadvantage of the transfer approach is the skill level and care required to produce the grammar and transfer rules. It can become very difficult to manage large sets of such rules, and the writers of such rules need to be highly skilled.

9.4.2.3 Interlingual Approach

The interlingual approach analyses the text into a language-independent representation called an 'inter-lingua'. From there target text can be generated directly. In terms of the data flow, this looks like the transfer approach, but with no transfer rules. Figure 9.4 shows a schematic representation of the data flow in the interlingual approach. Typically the analysis of the source text to produce the interlingua will involve a number of stages, usually a syntactic analysis followed by some kind of semantic analysis.

The interlingual approach is normally associated with 'knowledge-based translation', where the interlingua is used to represent the 'meaning' of the source text. However, there is no reason why an interlingual architecture has to use a knowledge-based approach and shallow interlingual systems are quite plausible which use little more than a syntactic analysis of the source text. Generally speaking, the deeper the analysis, the easier it is to write general translation rules.

Although there are no transfer rules to write in an interlingual system, the additional complexity of the analysis and generation components means that they tend not to be less expensive to produce than the transfer type of system.

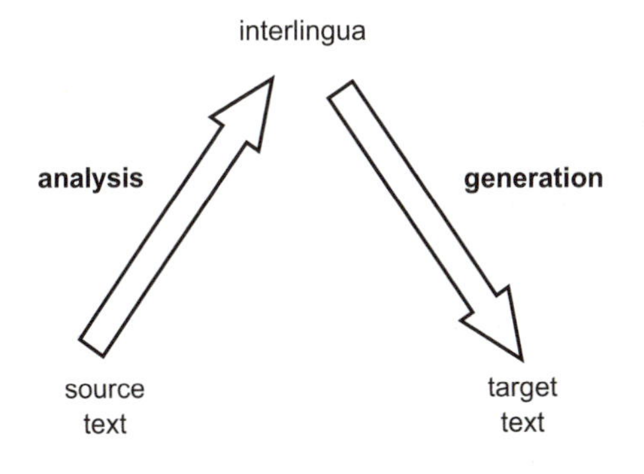

Fig 9.4 The translation process in an interlingual type translation system.

It has been argued that, ultimately, the only way to produce very high quality translations, is to try to incorporate human-like world knowledge into the translation system [19, 22]. However, the difficulties of doing this, for anything other than very narrow domains, make knowledge-based approaches which require hand-crafted rules very difficult to produce economically.

9.4.3 Machine Learning Approaches

Creating the large linguistic databases required by any of the machine-translation approaches described above clearly limits the coverage or accuracy that can be produced for a given development budget. It also makes the systems inflexible, in that it is very difficult for end users to carry out any customisation, beyond adding simple extensions to the vocabulary.

There is, however, a very large body of translated text. A variety of approaches are being investigated to try to make use of corpora of existing translations, rather than databases of hand-crafted rules [23]. The initial proposal of such an approach is most often attributed to Nagao [24].

Under the general rubric of 'example-based machine translation', there are a variety of techniques being investigated. One of the first systems to convince researchers that example-based approaches were worthy of investigation, was a system from IBM called Candide [25]. This system was intentionally very naive, and although the results did not then present a threat to hand-crafted techniques, they were very encouraging.

Most current approaches are hybrid approaches, typically making use of translation examples to refine some part of the translation system which is difficult to code by hand (normally the transfer rules) — see, for example, Microsoft's example-based system [26]. Microsoft claim to have the first translation system, making use of machine learning, that outperforms a hand-crafted system. However, the majority of their system is still hand-crafted.

The main difficulty of example-based approaches, is that their naivety means they are very poor at generalising. This in turn means that they require a vast amount of training data to produce adequate translations of arbitrary text.

9.4.4 BT Translation Research

Within BT, we are investigating an approach based on the reasonably unexplored technique of 'data-oriented translation'. The terms 'example-based' and 'data-oriented' may appear synonymous, but, in practice, data-oriented systems learn from examples which have had additional information provided by a human being. In our case, this additional information consists of monolingual analyses of the source and target texts in an example and a detailed alignment between the source and target analyses. An example of such an alignment is shown in Fig 9.5.

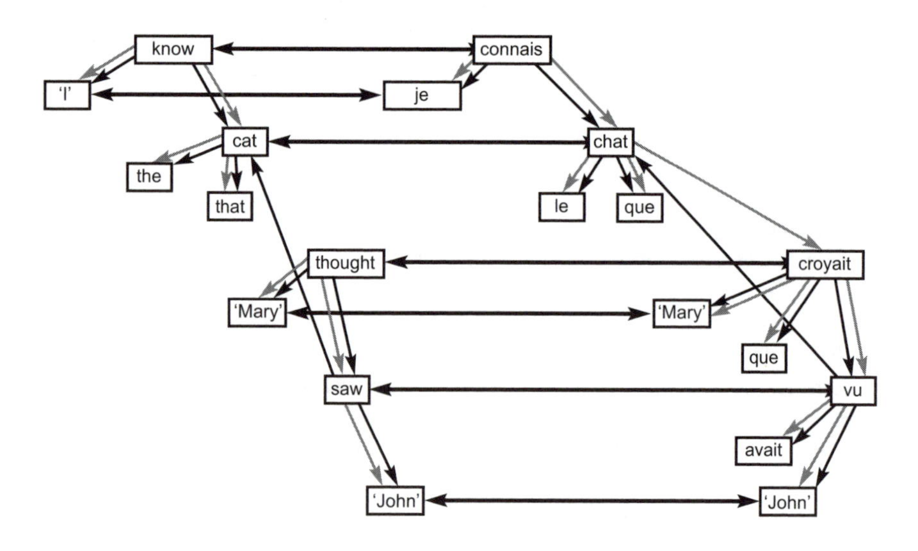

Fig 9.5 An analysed translation example.

From such analysed examples, it is possible to infer both source and target grammars and the transfer rules automatically. The skill level required to analyse translation examples is much less than that required to write the grammars and transfer rules manually.

The analysed examples act as a kind of bootstrapping mechanism, allowing the system to analyse certain examples automatically. The ability to learn from unanalysed examples will increase as the corpus of manually annotated examples increases.

We employ several generalisation mechanisms to be able to translate examples not previously presented to the system. As a first stage we break an analysed example down into small units (translation units) that can be reassembled to form new translations. In addition to this, there is a further generalisation process that creates new translation units by looking for regularities among the units derived from the analysed examples.

As the combination of small translation units in this way will typically produce many incorrect translations, there is a mechanism to infer constraints on which such units can be combined.

We have demonstrated our machine-learning algorithms on a small number of examples. Current research is focused on building up the number of examples to achieve a wide-coverage system and solving the scaling problems that arise from this.

9.5 Summary

If an application or service needs to be ported to other locales, then thought needs to be given to both the architecture and the low-level engineering of the system at a very early stage. In some applications, it may be a relatively simple matter of translating text on menus and in dialogue boxes, in others, it may mean managing partially dependent language and locale-specific databases.

Focusing particularly on the area of text display, we have seen some of the difficulties that can be encountered for languages that are not so well supported by international standards. In practice, *ad hoc* solutions are needed which may require significant compromises to be made.

We have also discussed the issues relating to automatic translation of text. Human-quality machine translation is still a very long way off, but useful approximate translations can be achieved. Since the potential financial rewards are so great, there is still much ongoing research in the area of machine translation.

References

1 AltaVista — http://www.altavista.com/

2 Systran — http://www.systranmt.com/

3 Yang, J.: '*Systran's approach*', The 1st Chinese Language Processing Workshop', CIS, University of Pennsylvania, Philadelphia (July 1998).

4 Yahoo — http://www.yahoo.com/

5 Logos — http://www.logos.com/

6 International Online Marketing — http://www.euromktg.com/

7 Mneimneh, H.: '*Interface design for the Arab Middle East*', Language International, **9**(3), pp 20—21 (1997).

8 WWW Consortium — http://www.w3.org/

9 Smith, W.: '*HTML goes multilingual*', Language International, **9**(4), pp 10-11 (1997).

10 Esselink, B.: '*A Practical Guide to Software Localization*', John Benjamins, Amsterdam (1989).

11 Rowe, T.: '*Asian type aesthetics*', Language International, **9**(1), pp 30-31 (1997).

12 Newton, J.: '*The Perkins Experience*', in Newton, J. (Ed): '*Computers in Translation*', pp 46-57, Routledge (1992).

13 Bernth, A.: '*Easy English: addressing structural ambiguity*', in Farwell, D. and Gerber, L. (Eds): '*Machine translation and the information soup*', Proc of 3rd Conf of Assoc for Machine Translation in the Americas, Springer, pp 164—173 (1998).

14 Bitstream — http://www.bitstream.com/

15 Microsoft — http://www.microsoft.com/

16 Adobe — http://www.adobe.com/

17 Arnold, D., Balkan, L., Humphreys, R. L., Meijer, S. and Sadler, L.: '*Machine Translation: An Introductory Guide*', Oxford, Blackwell (1994).

18 Nagao, M.: '*Machine Translation — How Far Can It Go?*', Oxford University Press (1989).

19 Nirenburg, S., Carbonell, J., Tomita, M. and Goodman, K.: '*Machine Translation: A Knowledge-based Approach*', San Mateo, Morgan Kaufmann (1992).

20 Halliday, M. A. K.: '*Machine Translation: An Introduction to Functional Grammar*', London, Arnold (1994).

21 McCord, M. and Bernth, A.: '*The LMT transformational system*', in Farwell, D. and Gerber, L. (Eds): '*Machine translation and the information soup*', Proc of 3rd Conf of Assoc for Machine Translation in the Americas, Springer, pp 344-355 (1998).

22 Goodman, K. and Nirenburg, S.: '*The KBMT Project: Machine Translation*', San Mateo, Morgan Kaufmann (1991).

23 Somers, H.: '*Review article: example-based machine translation*', Machine Translation, **14**(2), pp 113-158 (2001).

24 Sato, S. and Nagao, M.: '*Towards memory-based translation*', in Proceedings of COLING '90 (1990).

25 Berger, A., Brown, P., della Pietra, S. A., della Pietra, V. J., Gillett, J., Lafferty, J., Mercer, R., Printz, H. and Ures, L.: '*Candide system for machine translation*', in Human Language Technology: Proceedings of the ARPA Workshop on Speech and Natural Language (1994).

26 Menezes, A. and Richardson, S. D.: '*A best-first alignment algorithm for automatic extraction of transfer mappings from bilingual corpora*', in Proceedings of the Workshop on Data-driven Machine Translation at 39th Annual Meeting of the Association for Computational Linguistics, Toulouse, France, pp 39-46 (2001).

10

SMART TOOLS FOR CONTENT SYNTHESIS

M Russ and D Williams

10.1 Introduction

Before Gutenburg [1] and others made printing relatively cheap and easy, stories were related through story-tellers and travelling minstrels. The stories thus told were probably never the same twice. The art of the telling would have been to involve, interact and respond to the audience. This art has, to a large degree, been lost.

The most popular medium for story telling today is the television, and in television the story-teller does not know who is watching and has very little chance to involve or react to them. Television programmes such as *Big Brother* and *Pop Idol* both attempt to involve the audience and react to it through voting. These programmes have both proved very popular and are perhaps a testament to the audience's nascent desire to interact with a story line.

As broadband connectivity grows, and as television and the Internet continue to converge, new forms of such interactive, involving and personalised television will emerge. This chapter describes some early results from a set of prototype tools and an associated software architecture that, by using object-based media techniques, allows television programming to be personalised.

10.2 Understanding the Media Industry

The Nobel laureate economist Herbert A Simon has famously claimed that:

> 'What information consumes is rather obvious: it consumes the attention of its recipients. Hence a wealth of information creates a poverty of attention, and a need to allocate that attention efficiently among the overabundance of information sources that might consume it.' [2]

This encapsulates the problem that drives the media industry. The industry itself refers more prosaically to the need to attract 'eyeballs'. Eyeballs attract advertisers and it is the advertisers, at least for commercial TV, who pay for the production of the content.

There has been a huge increase in the amount of television that is broadcast, either through satellite, through terrestrial transmitters or through cables. In spite of this, the number of hours of television that an average viewer sees each week remains remarkable static [3] at about 25 hours per week. If eyeballs pay for TV, then how has this happened? Primarily it is through increasing revenues into the broadcasters (subscriptions in particular) and through a reduction in the cost of the production and transmission.

10.2.1 New Media for a New Millennium

While there are high levels of innovation in the device and technology fields it can be argued that innovation in the creative fields has not successfully transferred into the mainstream. There are many possibilities for innovative creation based on the abilities of local mass storage, broad-bandwidth connectivity and high-resolution output. But there are huge economic risks associated with attempting to develop a new creative format and the dominant incumbents have a vested interest in the *status quo*.

Formats such as cinema started with small-scale innovative operations working from a suitcase, such as the Lumiere brothers [4]. We might expect innovation in this converged world to begin from dynamic, innovative and daring creatives who operate outside the mainstream media business. There are a range of media artists and academics experimenting with new forms of film and cinema creation and trying to extend the natural limits of film and video to take advantage of the new capabilities that converged media can offer.

10.2.2 A Key Opportunity for Broadband

Broadband connectivity offers the potential for more than merely broadcasting television content. It would be folly to use it to copy the offers from satellite or traditional broadcasting. Instead, it should look to exploit its key attribute — that the service provider knows who is watching and can respond to their individual needs and preferences.

At a simple level this personalisation can be achieved by preselecting content that meets the needs of the viewer. If this preselection is done at a programme level, this functionality is merely equivalent to that of a personal digital recorder. Allowing an individual programme to be flexed to suit the preferences of the viewer requires control at the 'scene' level. It is not trivial. There are key challenges in the creation of attractive content that responds robustly to this type of 'flexing' without losing its narrative sense. In addition the management and delivery of this new form of multimedia content will be unlike any other. Think for example of the issue of the management of content rights which only become payable when a particular scene is played and that has to be monitored on a per viewer level!

10.2.3 An Approach to New Media Production

The problem with trying to describe a new multimedia content genre is that people are understandably unfamiliar with things they have not seen before! An inventor of television meeting radio company executives in the first quarter of the last century would not have been able to describe the immersive power of *Inspector Morse* or the compulsive viewing that is associated with *Big Brother*. The description would necessarily have been limited to using metaphors like 'Radio with Pictures', which is a very inadequate way of conveying the appeal and magic of television.

A critical aspect of the new genre is the ability to flex the content to control the version seen by the viewer. This is a new form of freedom for anyone who has grown up in a world where media is in the form of audio or videotape, CDs, DVDs or any other form of media where the implied delivery mechanism is one of streaming in linear time.

For the new genre, the end product of the production process is not a single piece of shared media content. It is a collection of well-described media fragments (media objects) together with a set of constraints and relationships that will allow them to be assembled into a coherent finished piece of media. In a way that is strongly analogous to software objects, this 'smartening' of content adds methods and functionality to the merely descriptive metadata for the content [5].

The 'patent-applied for' tools, that have been created for this new 'Smart Content', support and enhance the video editing and production process. They bring together the roles of logging, editing and producing into one repository of the rules, glue, hints and tips that would normally only have a transitory existence during the creation of the piece of media. By capturing this normally ephemeral information, reuse is facilitated both in the production process and in post-production. And by presenting all of the information in context, the tools enhance the production of the immediate work-in-progress, as well as its potential future reuse. The ability to capture this otherwise transitory, but clearly valuable, information provides a powerful incentive to use the tools to add meta-data, not at a per programme level (as is current practice) but at a per scene level. It also sets the capabilities of these tools apart from other media mark-up tools.

Using the reality TV programme *Big Brother*, we can begin to glimpse what might be possible using what we have called 'Smart Content' production principles. Under current production constraints the footage from the numerous cameras around the *Big Brother* household is logged and edited together to create the requisite two 30-minute programmes — one family-oriented 'tea-time' version, and one adult-oriented 'post-watershed' version.

In this process the editor has exerted considerable artistic skill and utilised both previous experience and considered judgement. The editor's constraints and objectives include:

- the creation of a programme of the correct length;
- the accommodation of advertisement breaks at appropriate intervals;
- the creation of a programme that exhibits the 'house-style' that defines *Big Brother*;
- the creation of a programme that is seen to be fair to all the contestants;
- the creation of a programme that complies with national decency conventions and that will not offend its audience;
- the creation of a programme with appropriate aesthetic appeal.

Choosing appropriate material from the stock generated within the *Big Brother* house is a formidable task but it is well executed and the format and concept of *Big Brother* remains highly appealing — hence many millions of people watch it.

However, if the marking-up techniques described above were adopted, then each video clip would be logged (as it is already), and the ephemeral information, such as the dependency between two particular clips would be captured, not lost. This would enable multiple versions of the *Big Brother* programmes to be produced at very low additional cost. These versions could focus on a particular individual, they could focus on a particular room, a particular activity or, condensing material from many days viewing, they could concentrate on behaviours exhibited when drunk, or on 'nomination days'. The possibilities are endless, and are made possible merely by marking-up once and reusing many times.

Which of the possible combinations (if any) would be attractive to a viewing audience is a moot point and it would take the creative inspiration of a television production house to determine a concept that 'built in' these possibilities from the storyboard stage to make it compelling. This sort of television is not (or need not be) just passive 'lean back' TV. It could be 'lean forward' TV in which the viewer is invited to make involving choices that allow the deeper exploration of the captured media. It will challenge the programme-maker's art, as the enjoyment for the viewer could come as much from the process of investigating the material as it does from viewing the end product. The journey, not the destination, could be the goal.

10.2.4 New Tools for the New Multimedia

To date, we have used the prototype mentioned above to reverse-engineer traditional television programmes. This is far from ideal. Original material created from the outset with flexible production in mind would be preferable — and more constructive.

The architecture in Fig 10.1 is used to create the personalised media.

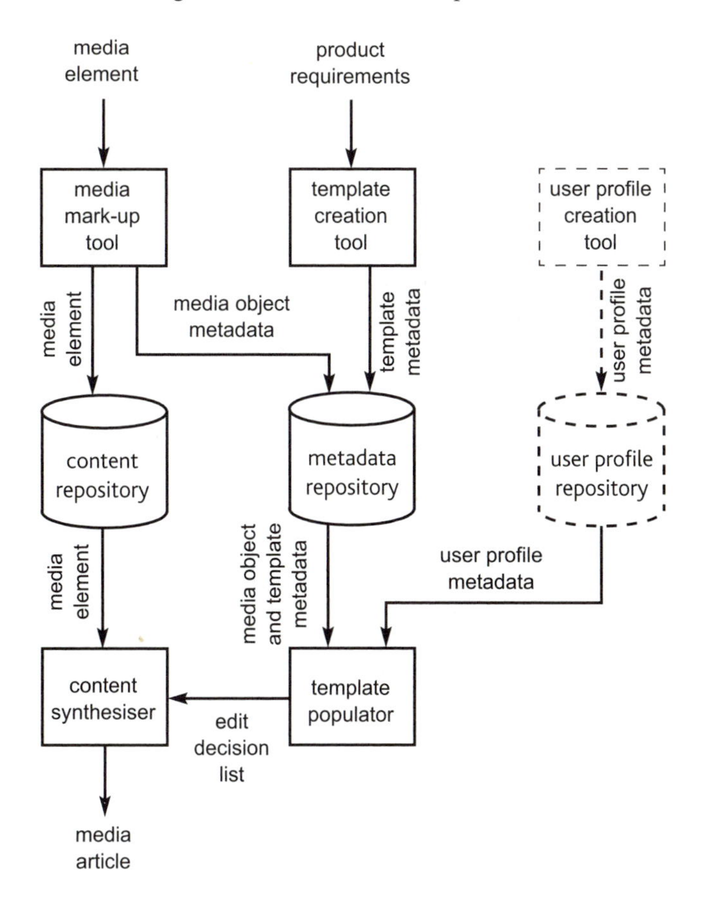

Fig 10.1 Schematic overview of the architecture elements used to create personalised TV programming.

Media clips are stored in the content repository, while the information that describes the clips (the metadata) is stored in the metadata repository (an object-oriented database). The idea of the 'content repository' or 'Media Asset Management System' is well established. However, in this work the output from the production and editing phase is a collection of well-described media fragments, together with a set of constraints and relationships. In conjunction with a suitable means of resolving the dependencies, this will allow the media fragments to be assembled into a coherent finished piece of media. The process flow of the tools is in four stages:

- media mark-up;
- template creation;
- template population;
- content synthesis.

The process also depends upon the existence of 'user profiles'.

10.2.4.1 Media Mark-up

In this stage, the media clips are annotated with structured metadata which is intended to describe the content with a hierarchy that leads from physical attributes to abstract meaning. This description stage includes a range of metadata tags ranging from basic low-level data (such as filename, size, etc) through to high-level metadata that identifies the media object in terms of its aesthetic and narrative status within the canon of media clips.

The data model used includes the data types shown in Table 10.1.

Table 10.1 Data types used in the model.

Level	Data type	Examples of the data type
High	Relationship	This data helps establish which clips are similar, which form a natural sequence that should be played together and which clips can be related through some dependency
	Domain specific	There are extensible sets of parameters that help to identify the clip within the specific context, e.g. in the *Big Brother* example 'Nomination day' and 'Eviction day' could be domain-specific attributes
	Conceptual	The (subjective) importance to the plot, the role the scene plays in the narrative, its aesthetic appeal, etc
	Structural	The actors involved, dialogue, type of action (fight, fright, etc)
	Media	The video format, duration of the clip
	Usage	The content owner, rights payable, etc
Low	Creation	The date of creation, title, etc

The lower level metadata fields were taken from MPEG7 standard, while the definition of the more subjective elements (domain specific and relationship data) are an evolving product of the current work. To accommodate this evolution the data model has been designed to be flexible to accommodate the emerging demands on the description of the media.

It is also necessary to attempt to identify and record relationships that exist between media clips. Capturing and then acting upon this relationship information is very important if different versions of the media are to have any narrative sense. The tool for capturing this complex relationship metadata was designed to be wholly graphical. The media clips (represented as thumbnails) were grouped together with associated iconography to indicate characteristics such as whether the relationship between the clips was causal, sequential or an indication that the narrative effect of the clips was equivalent.

10.2.4.2 Template Creation

Templates help the editor or producer to define the format of the finished media. A template does not, in itself, define the content of the programmes but defines the structure or format that the programme takes. This is perhaps most easily recognised with programmes like the News, whose format is familiar to many (see Fig 10.2).

The content within each section of the template needs to be defined. It is in this definition stage that the opportunity exists to personalise the programming. The news editor chooses the most important news story using some subjective measure of importance, attempting to identify the news story that, to most of the intended audience would be regarded as the top story of the day. It must be acknowledged that the role of the news editor on a national TV channel takes on a significant cultural role.

The editor's ability to affect the attitudes of a nation bring with it very serious responsibilities, including providing a balanced view of the subject matter. However, the cultural and sociological effects of the technique we describe will not be addressed here.

This 'editor-driven' process for choosing content could be replaced with a 'viewer-driven' process. This can only be accomplished using the new multimedia genre described here — remember that this is not intended to be a broadcast medium in the conventional sense.

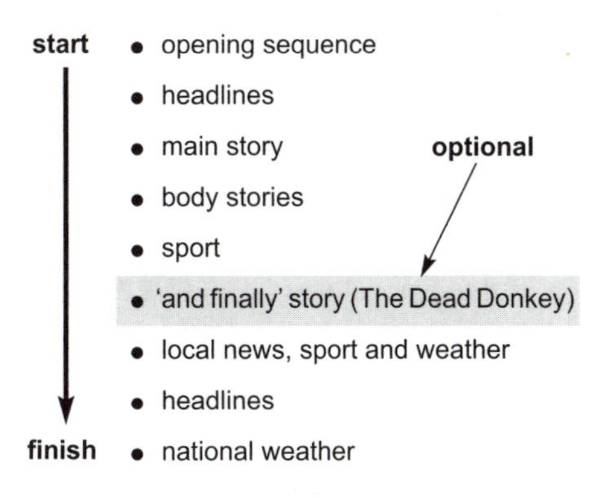

Fig 10.2 The 'News' template.

With the proposed system each news story would be marked up with a range of associated descriptors, which may include metadata tags such as financial, foreign, social, sport, local, general interest or celebrity.

10.2.4.3 User Profile

Simple user profiles, based on XML schemas, have been used to enable a viewer to express their personal preferences. The user profile is key in enabling the viewer to receive a personalised television programme. In the example of the news, the profile could indicate whether foreign or national news is of most interest, which regional news is of most interest and in which sports or sports teams a viewer is most interested.

For areas where viewers are unable to express their preferences unambiguously, or where clear descriptors are not feasible, then associative or referential methods can be used instead of more direct naming conventions. For example, while people might not be able to agree about exactly the point at which a red colour might be better described as an orange, this does not prevent people from using colours. Tins of paint often use unusual, antiquated or obscure words to try to describe the colour, but often the most useful indicator is the strip of colour next to the words, or the actual paint itself. Thus, while viewers may not realise that the generally positive and light-hearted stories often found at the end of the news may be referred to as ‘dead donkey’ stories, they will understand that they enjoy ‘stories like that’.

10.2.4.4 Template Population

Template population is the term used to describe the assembly of the content fragments (media clips) into a coherent form, using the structural information from the template. Queries in each section of the template define which stories should be retrieved to populate that part of the template. If queries are structured such that they request media clips that best match a user profile then, although a generic template has been used, each viewer will see a potentially unique programme.

A personalised template for the news could stipulate that the stories appear in priority order as defined by the user profile. Thus for one viewer foreign news could appear above national news stories, for another viewer it would be vice versa. In addition, the selection of the regional news could be driven by the user profile (and not necessarily just driven by the viewer’s location) and similarly for the sports news. Another example of this type of reusable generic template would be in a programme where the format has a formulaic structure, e.g. a ‘makeover’ programme. Each week, a location is shown in its original state, then a team of artisans accomplish the changes required to improve or enhance the location, and then the original and revised location are compared. The template for each programme is the same: only the clips of media change. A user profile might be used here to control the exact clips that are chosen, so that a specific presenter or artisan would be used in preference.

The concept of personalisation using a template is familiar to the world of the Internet where personalised Web pages bring viewer-driven content to the page. The template structure for a Web site is one of page layout templates. The TV programme requires a different 'syntax' to define its boundaries, and clever processes to resolve complex constraints that could arise from relatively fuzzy queries.

During template population the queries in the template are executed against the information about the available media content. The template populator actions the queries in the template, reconstructs the relationships and resolves constraints to create an edit decision list (EDL). The EDL is a set of instructions that dictate in which order the various media clips need to be played to enact the 'viewer's cut'. The result of this process is an EDL for a programme consistent with the overall structure of the template.

This is not trivial. For example, returning to the example of the News it is possible that there are conflicts between the template structure and the query. A viewer's declared preference may be for foreign news and the template populator may retrieve 20 minutes of foreign news stories. However, if there is global constraint such as 'the news should last no more then 15 minutes' then clearly there is a conflict that needs to be resolved. Additional conceptual metadata such as 'importance' could help to resolve this issue. In media forms with a more traditional narrative there will be even more complex queries to resolve that will lean more heavily on the relationship data that was added at the mark-up stage.

10.2.4.5 Content Synthesiser

The content synthesiser is the final element in the production chain, and is responsible for generating the audio and video for the viewer. The content synthesiser is able to play media clips from locations specified within the EDL (these clips being either locally-stored or streamed over a network), and support a variety of popular coding standards for PC audio and video. The content synthesiser links together the media clips so that they appear to be a single piece of video — the 'behind-the-scenes' assembly of the individual media clips 'on-the-fly' is not apparent to the viewer of the final output. Additionally, the content synthesiser is required to combine separate clips, providing transition effects between the segments, and to offer facilities such as picture-in-picture viewing.

The content synthesiser should be a portable, lightweight application that can be deployed at the edge of the network, or within the consumer's device (such as a PC, a games console like the PS2 or X-Box, a set-top box for cable or satellite TV, or a personal digital recorder like TiVo). This means that TV-like programming can be synthesised at the point of consumption, with considerable benefits in terms of the level of personalisation that can be achieved.

10.3 Experiments in the Creation of Personalised Television

The capabilities of the software have only been tested with programming material intended for traditional editing treatment. Nevertheless this has tested all the major features of the software and led to some important subjective observations about its capabilities. Programming sources as diverse as feature films, home movies and reality TV programmes have been tackled. In each case the canon of media content amounts to no more than 30 minutes footage in total.

Templates have been created and, by use of different queries within the templates, different customised versions of the content have been created. The development of templates is ongoing, with a number of enhancements planned to increase their functionality.

A number of people worked with the tools — all were very familiar with the intention of the project, but the level of nonlinear editing skills that the users possessed varied from beginners to a professional editor. The programmers who have written the software code have not been used to test the tools, preferring instead to keep a level of ignorance between the users and the creators in order that the tools are tested fairly.

10.4 Results

10.4.1 Media Mark-up

In reverse-engineering the content, it is necessary to 'cut up' the programmes into short sections or scenes — which we have called 'media clips' in this chapter. These scenes were then marked up and templates were created to allow different versions of the original programme to be created by flexing some of the parameters within the template queries.

This task requires a decision about the ideal duration of a media clip. If the clips are too long then there is very little variation possible in customised versions. If the clips are too short then the task of adding the metadata becomes too onerous. For the scale of project tackled to date media clips vary in length from 5—30 seconds. This, however, is primarily a function of the source content and of the editor's subjective opinion of an appropriate instant to cut up the media.

It is well known that there are differences in scene length in TV programming and that the choice of scene length can bring with it a great change in the pace of a programme. No absolute optimum length can be prescribed. It will always depend on the nature of the material.

While media clip length will not be a golden rule that drives the process, an anticipation of the ways in which the final output will be flexed will be a key driver dictating the nature of each media clip. Sufficient granularity must be created to be able to offer clear separation of versions based on the type of versioning that will be made available to the viewer.

The ability to capture relationships between different media clips is an essential part of the tool. This allows the editor to indicate particular dependencies (such as if this clip is shown then this one must also be shown) and relationships (such as these clips all have the same effect on the narrative so any of these may be used interchangeably). The use of a graphical metaphor, with simple iconography, to capture this more complex relationship data was found to be very successful with test users quickly appreciating the 'grammar' of the iconography. This method of capturing relationship data has been essential in maintaining a comprehensible narrative in the finished versions. The relationships introduce a number of complex constraints that have had to be resolved using recursive operation between the template populator and the database of media objects.

A current limitation in the process is that audio and video are edited together. This introduces some very staccato edits between media clips, thus making them very noticeable. Separating out the audio and video editing and using stock editing tricks (fades, dissolves, etc) will probably solve this, though it will make the process more complex.

10.4.2 Templates

The template concept has proved easy for most people with editing knowledge to comprehend. However, the way in which the application was built relied on highly logical expression of the queries using classical Boolean logic strings. While the precision this offers is absolutely necessary from the point of view of the software programming, the sense of the AND and OR operators in a language like English and Boolean logic are not the same.

Using the English language, in reference to a table full of pool balls, the phrase 'pass me the red and yellow balls' would normally be interpreted as 'pass me all the balls that are red and pass me all the balls that are yellow'. The interpretation is actually fulfilling the Boolean query 'pass me the red OR yellow balls'. The Boolean query 'pass me the red AND yellow balls' would return no balls, as no pool balls are both red and yellow (Fig 10.3)!

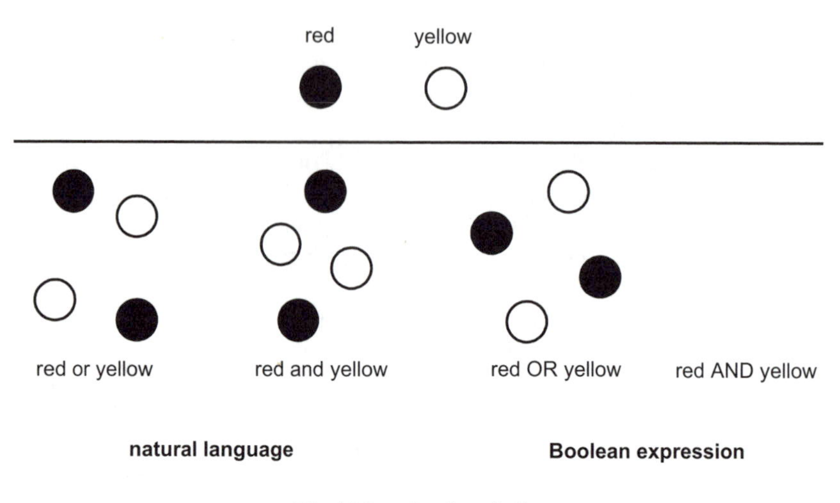

Fig 10.3 Boolean balls.

The Boolean expression is precise. However, the users of the tools are unlikely to speak or think in terms of a mathematical abstraction like Boolean logic. It is much more likely that they will only speak or think in terms of a natural language. In subsequent versions of the software, abstractions will be made from the Boolean terminology in order to give the users of the tools an intuitive and unambiguous interface.

10.4.3 Content Synthesiser

The capability provided by the content synthesiser is, at first glance, comparable to an application that utilises the power of readily available video-playing components like Windows Media Player or Apple Quicktime. Both of these components were evaluated and rejected — the complete Windows Media Player being too inflexible, while the documentation for the Quicktime player seemed to be aimed at Macintosh developers.

The content synthesiser was built using elements of Microsoft's DirectX 8 technology (which is used to create the stand-alone media player). It was written in Visual C++ and built as a Windows ActiveX control. This enabled (as required) the content synthesiser to be a portable and lightweight application that could be reused in other tools that require a media player, or even to be dropped on to the Windows Active Desktop, to provide video as a background to other Windows applications.

The content synthesiser has been used with a range of media sources, successfully playing an edit decision list of over 100 media clips. It has also been used with media files as large as 500 Mbytes.

10.5 Next Steps

The functionality of the tools described above is a necessary step in the creation of personalised programming; but several technical, artistic and business challenges remain. These include:

- understanding which of the dependencies and rules that control the art of film-making can be parameterised;
- making the addition of metadata as intuitive as possible (or even as automatic);
- stimulating the production of content that is intended to be treated as media objects that can be assembled to create customised instances of the media set.

By working together with media artists and innovative media creators, these problems will be solved and more detail will become visible on the vague outline of a new media genre. The ongoing development of these tools is a key part of this exploration.

Work so far has concentrated on video media. As mentioned in other publications on this technique, the concept can also be applied to the production of other media material, specifically computer games. In common with TV programming, computer games are built from a range of media assets, which similarly are built from small fragments of content. The next steps include:

- producing the right data models to suit specific genres of game;
- producing tools that suit the production process of games;
- converting a games engine into a game content synthesiser.

10.6 Summary

Media delivery is changing and opportunities for new media genres based on the creation of personalised television programming emerge from the availability of broadband connections, low-cost mass storage and the adoption of object-based media production techniques. An analysis of the issues involved in the creation of personalised media has been completed. This suggests that the output of the production process should be a collection of well-described media fragments together with a set of constraints and relationships that will allow the media fragments to be assembled into a coherent finished piece of media. A set of tools has been developed that allow this production end-point to be reached.

Early content experiments based on reverse engineering of traditional television programming have confirmed that the developed software can deliver rudimentary personalised TV-like programmes. More work is required to develop content that exploits the full capabilities of the tools through adopting the goal of personalised programming from the outset of production process.

References

1 '*The British Library Guide to Printing*', British Library Publishing (November 1998).

2 Herbert, S.: '*Designing Organisations for an Information-rich World*', in Greenberger, M. (Ed): '*Computers, Communications and the Public Interest*', The John Hopkins Press, pp 40-41 (1971).

3 Broadcasters Audience Research Board (BARB) — http://www.barb.co.uk

4 Rittard-Hutinet, J. (Ed): '*Auguste and Louis Lumiére — Letters*', Faber and Faber, London, p 302 (1995).

5 Russ, M., Kegel, I. and Stentiford, F.: '*Smart Realisation: Delivering Content Smartly*' Journal of the IBTE, **2**(4) (October-December 2001).

11

PERSONALISED ADVERTISING — EXPLOITING THE DISTRIBUTED USER PROFILE

G Bilchev and D Marston

11.1 Introduction

Although the concept of a user profile is well understood and well used on the Internet today, we start from a different assumption — each user already has many profiles held by the service providers with whom the user has been interacting. The overall user profile (called from here on the 'distributed user profile'), therefore, is the set of all of these distributed profiles.

The problem with the distributed user profile is that it may not be easy to use. For example, consider the situation where one user has two different profiles with two different providers (held in two different databases) with no link between the two profiles. If privacy was not a concern, and if the user has used the same identity with both providers, then, in principle, the two profiles could be consolidated, but, in the real world, two separate organisations will be obliged by privacy protection acts not to disclose user information. Two possible exceptions to the above arguments would be in cases where:

- one organisation merges or buys another;
- explicit permission is asked from each user as to what information can be shared with which other organisations.

An example of the former would be the purchase of Abacus by DoubleClick [1], which spurred a lot of privacy controversy, and an example of the latter would be Microsoft's Passport service [2], where users are asked what information can be shared with sites participating in Passport.

In general, however, it is safe to conclude that the distributed user profile is fragmented due to the lack of links between the distributed components comprising the profile.

11.2 On-Line Personalised Advertising — Overview

One ideal application for the concept of a distributed user profile is in on-line personalised advertising. Unlike many other personalised advertising systems, which focus on how best to build a user profile from browsing behaviour, an advertising system based on distributed user profiles builds on the assumption that each user already has many profiles held by the service providers with whom they have been interacting.

The rationale behind this initial assumption is that numerous high-street retailers have been running loyalty programmes where data about specific users has been gathered, and therefore, detailed shopping profiles have been built (or can be built). These shopping profiles include both temporal patterns (when the user shops) and location patterns (at which geographical locations the user shops). The same is true for on-line retailers with whom the user has been interacting on a recurrent basis and hence a shopping history exists that can be used to build a profile. An example would be the recommendations service provided by Amazon, which can be viewed whenever a registered user visits Amazon's site (see Chapter 15 for more details).

The goal of a personalised advertising system based on the distributed user profile will therefore be to create the links between the various distributed profile components in a manner that preserves or respects user privacy. This means that an organisation that was unaware of a certain aspect of the user's distributed profile should not be able to deduce it by simply using the personalised advertising system. As an example, consider an on-line advertising broker serving a personalised advertisement from a grocery store. Based on the past shopping patterns of a user, the grocery store might conclude that the user will be interested in a new range of baby food, but this information should not be visible to the advertising broker. In implementations where the broker serves the advertisements, the broker should have a privacy policy not to store the generated banners or use any means to extract information from the images and link it to users.

11.3 Models of On-Line Advertising

Before describing the essence of the distributed personalised advertising system, it is worth noting the existing models/architectures of on-line advertising.

11.3.1 The Broker Model

The broker model consists of an advertising broker running a network of participating publishers (Web sites that would display the advertising banners) and advertisers (entities that have advertising banners to display), as seen in Fig 11.1. Usually, apart from hosting the advertisement-serving software, the broker will provide some kind of targeting criteria. These can be by geographical location, by publisher site, or by more detailed user profile (age, gender, browsing interests and patterns). Often the targeting is priced per impression (a measure of on-line advertising success, see section 11.5 for more information), and advertisers can plan their campaign using value-added tools from the broker. The broker takes a share of the campaign money and passes the rest to the publishers.

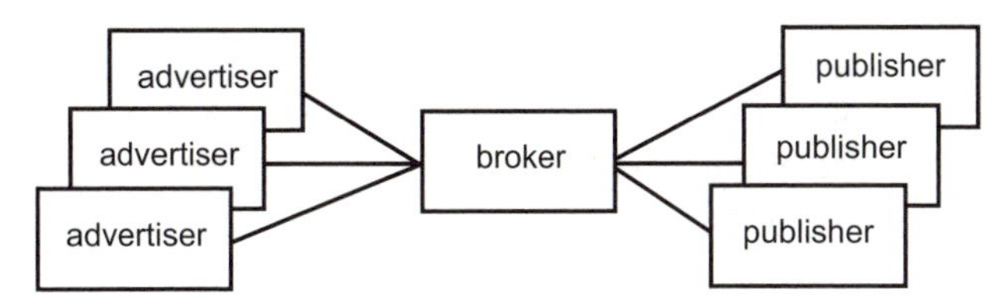

Fig 11.1 Broker model of on-line advertising.

11.3.2 The Portal Model

The above model works well for small and medium-sized publishers, but large publishers (such as Yahoo and BT Openworld) are in a position to offer advertisement brokering themselves. They usually do this by hosting the advertising platform (Fig 11.2) and managing their own network of advertisers. The advertisement serving platforms are often the same platforms as in the broker model, but licensed as stand-alone software hosted by the publishers. The advantage for the publisher is that the user profiles are held within their platform rather than the broker's. This is often important to the portal due to the perceived value of 'owning' the customer (or their profile).

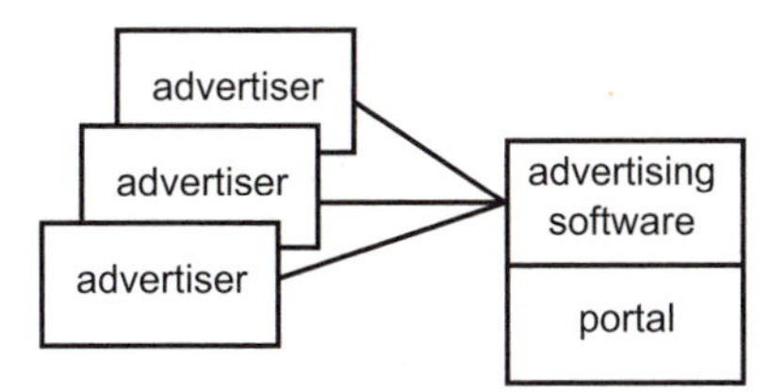

Fig 11.2 Large portal model of on-line advertising.

11.3.3 The Advertiser Model

Sometimes large, on-line shops or high-street chains might decide to advertise directly to their customers. In cases where they know the customer's e-mail address or mobile number, they may use these channels to send recommendations via e-mail or mobile text messaging. As e-mail and messaging channels are considered quite interruptive, what is more interesting with regard to on-line advertising, and this chapter's focus, is that it is also possible to use banner advertisements to essentially provide the same personalised recommendations. The key here is to be able to recognise the user and dynamically generate a banner with a specific recommendation.

In this model the advertiser is managing the relationship with all the publishers (Fig 11.3), and it is therefore also quite possible that the publisher is not paid by number of impressions, but by generated revenue. An example is the Amazon Associates programme [3], where a publisher gets a percentage cut of the value of the generated purchases when the user has followed a link from the publisher's Web site into Amazon's Web site.

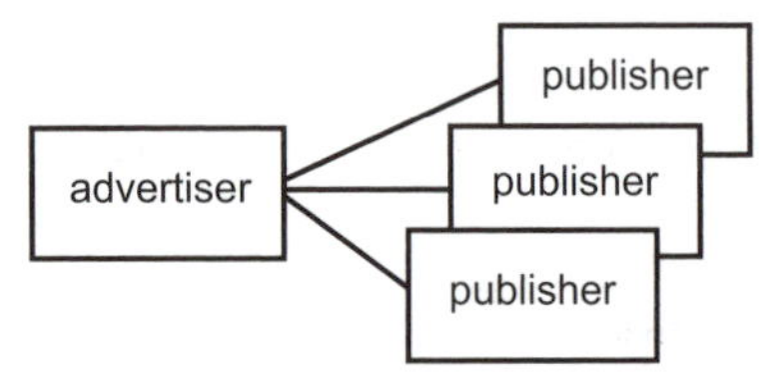

Fig 11.3 Advertiser model of on-line advertising.

11.3.4 Relative Merits of Using a Broker

The advantages of using a broker from an advertiser's point of view are that the broker will manage the relationships with all publishers. The publisher's network will usually be larger than a single portal and the broker will also offer value added services such as profiling of the publisher's sites. The broker could also have a better profile of the user, since many publishers would have used the same broker and the user's history file (sometimes called a reputation) could have been built more comprehensively.

The advantages of using a broker, from a portal's point of view, are access to a greater network of advertisers, and broker management of the relationships with all the advertisers.

11.4 Distributed Profile Advertising

11.4.1 Phase One — Distributed User Profile Creation

Distributed profile advertising is a two-phase process. The first phase is to actually create the distributed user profile. The creation process consists of creating links between the various user profiles rather than adding more data into a single profile (Fig 11.4). When creating the links, no user-identifiable information should be exchanged, as this would violate the privacy design principles. An advertiser that knows the user as 'George' would create a new identity for that user to be shared with the advertising broker, but this newly shared identity should be meaningless to any party other than the advertiser. The broker would then link the new shared identity with the identity by which it already recognises that user (or browser). The same process will take place with other advertisers, and therefore, at the end, the broker will have a set of shared identities mapped to its own identity for the user in question.

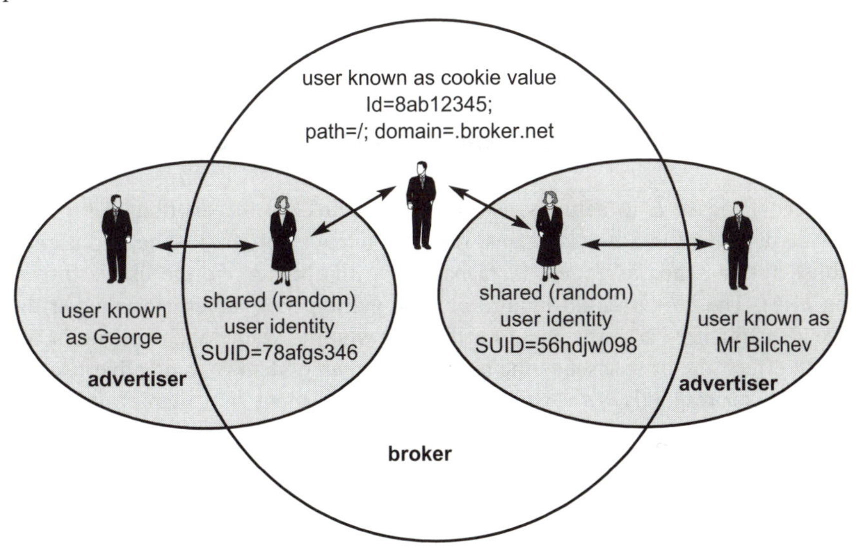

Fig 11.4 Linking the distributed user profiles.

Figure 11.5 shows one possible implementation of 'phase one'. When the user visits the advertiser's Web site (and the user is recognised by either logging in or by

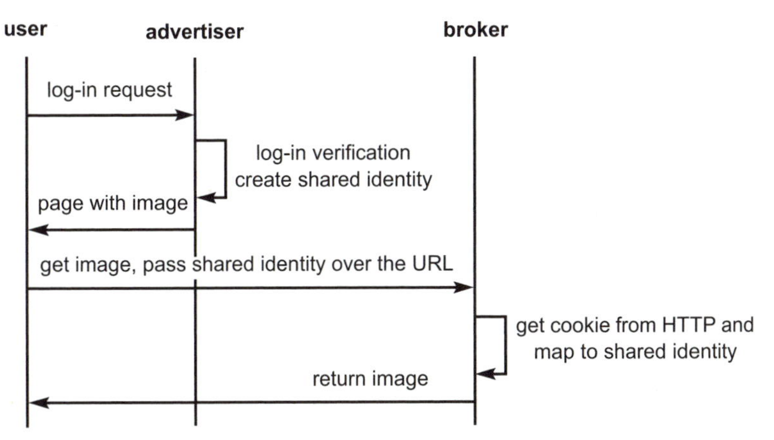

Fig 11.5 'Phase one' of distributed personalised advertising.

a persistent cookie), the advertiser includes in the returned Web page an image coming from the broker's domain. The image URL will carry the newly generated shared identity for the user in question as a parameter. When the broker receives the request, it will read the shared identity and its own cookie coming from the browser, and will map the shared identity to that cookie. This completes the process of creating the link between the profile held at the advertiser and the profile/reputation held at the broker. This would be repeated once for each participating advertiser, so phase one is a multi-step process, but needs to be executed only once for each participating advertiser, and does not require any input from the user.

11.4.2 Phase Two — Distributed User Profile Access

The second phase is to actually use the created links in the distributed profile to provide personalised advertisements, recommendations or coupons to end users. A publisher first sends a request for a banner to the broker via the user's browser (Fig 11.6). The broker could inquire of each participating advertiser whether they have a personalised advertisement for the user who has visited the publisher's Web site. In effect, the broker sends the appropriate shared identity to an advertiser and the advertiser responds with 'yes' (an offer/advertisement is available) or 'no' (no

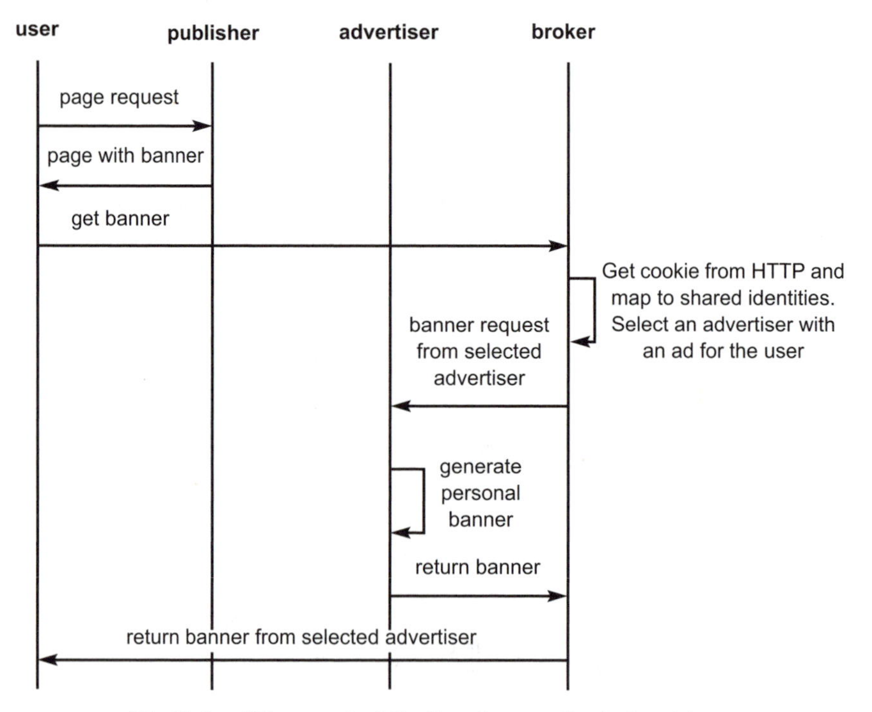

Fig 11.6 'Phase two' of distributed personalised advertising.

advertisement is available). (This process could also be executed in advance by each advertiser flagging those users for whom they have available offers.) After selecting an advertiser, the broker then requests the personalised advertisement, and returns it to the user.

The ability to target individual users could create an interesting market for on-line advertisements. Users who respond more to on-line advertising and with larger budgets to spend could become more attractive to certain advertisers. This could create an auction-like scenario for eBanner advertising where advertisers could attribute value to the banners they generate and bid for the individual user's attention through the broker.

11.4.3 Example of Distributed Profile Advertising Application

The effect of the distributed profile advertising system can be seen in Fig 11.7, which shows an advertising banner at the top right of the page giving a recommendation that fits the interests of the user. The same recommendation could be viewed from the recommendation section of the advertiser's site, but this would require the user to visit the advertiser's Web site. The distributed profile advertising system allows such recommendations to reach the user even when the user is not

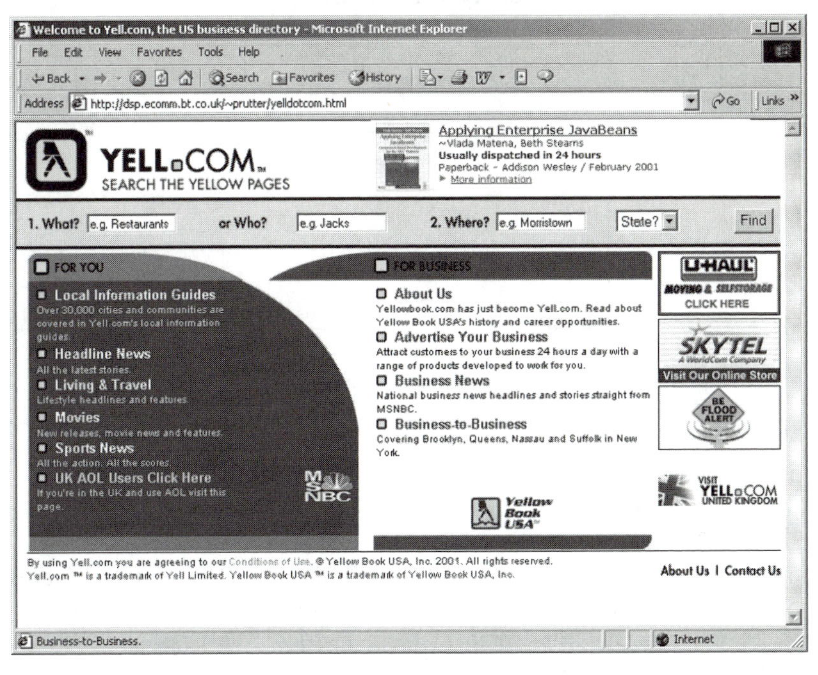

Fig 11.7 The recommendation banner demonstrates distributed profile advertising in action.

browsing the advertiser's Web site, but visiting a publisher from the broker's network.

It is interesting to note that the distributed profile advertising system works well with off-line advertisers. For example, a grocery store that has been using a loyalty card system to build detailed shopping profiles may decide to use the distributed profile system to generate personalised offer coupons for on-line users, who would then print them out (or send them to their mobile telephones) and visit the store. In this scenario, part of 'phase one' will be executed off-line with the help of the purchasing receipt (Fig 11.8). The offer/claim code, found on the receipt, is related to the shared identity, so that, when the offer code is used to claim the offer on-line, the remainder of phase one can be executed as before.

This chapter does not consider technologies for the actual profile compilation, offer recommendation or dynamic generation of personalised banners. These are considered to be available and deployed within the advertisers' domain.

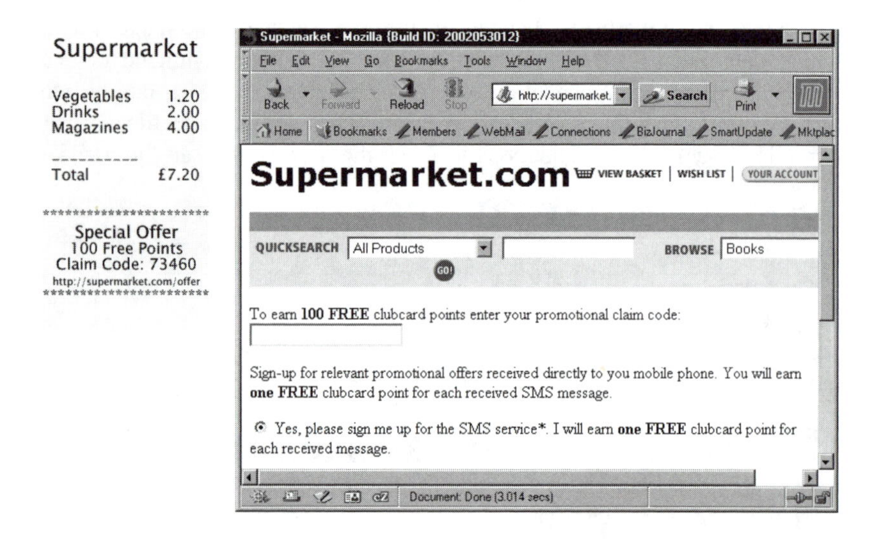

Fig 11.8 Using a purchase receipt to create the distributed profile.

11.5 Performance Metrics for Distributed Profile Advertising

Performance metrics are required to quantify the effect of on-line advertising campaigns, and, in examining the value of the distributed profile advertising system, it will be useful to describe some of these metrics. The usual parameters that describe the success or failure of an on-line advertising campaign include total reach, advertisement impressions, click-through and generated revenue.

Total reach for a campaign is the number of unique users that have been exposed to the on-line banner(s), but not necessarily seen it (nor paid any attention to it). The total reach for a site is either the number of users per month or the number of unique users per month that visit the site. The total reach of a site is a feature that could attract advertisers to use that site, but there are other features that become relevant for targeted campaigns such as gender, age, location, and income bracket of the site visitors, to name a few. Advanced advertisement brokers would offer site profiling tools [4, 5] to help an advertiser match their campaigns to relevant Web sites.

Advertisement impressions (usually measured in thousands) reflect the number of times an advertisement has been exposed to users. It does not measure any interaction and does not even guarantee that the user has seen the advertisement. Since the banner space on a portal is a resource, which is being sold during an advertising campaign, the publishers are quite happy with this measure as it reflects utilisation of their Web page space. Advertisers would, however, prefer some sort of a measure of advertisements that have been actually seen — as described below.

Click-through is a measure of the interaction between a user and a banner. Sometimes this measure could also include mouse-over events, but normally it reflects the number of users who have clicked on a banner.

Generated revenue is another measure that can be used to attribute value to on-line advertising campaigns. It has not been very popular for a number of reasons, not least that publishers would prefer to generate revenue per used resource (i.e. banner space). Also, generated revenue does not measure the effect of brand building. Without personalised banner advertisements, generated revenue would not be very effective, as users are less likely to proceed to a purchase of an item that is not relevant to them. Generated revenue, though, is very well suited for the advertiser model (section 11.3.3), where publishers participate in an associate programme and receive a cut from the generated purchases.

The performance metrics described above could also be used in measuring the success of the distributed profile advertising system. In terms of total reach, the important factor here is to measure the number of users that have profiles with the advertiser as opposed to total number of users. For example, an advertiser who is a high-street retailer in the UK will be more interested in advertising on a site that receives hits from UK shoppers who visit the retailer, and use a loyalty card. Also, as the advertisers know the shopping profile of the users, the value of an impression would increase. The argument is that advertisers could selectively target higher spending users. The distributed profile advertising system should also increase the click-through rate, as the personalised banners will be more relevant to users.

11.6 Privacy in Advertising

When investigating the design and implementation of any form of personalised advertising on the Internet it is important to consider the privacy implications of that advertising, and to do this it is important to understand what we mean by privacy, and why users are concerned. Privacy is often defined as the ability to determine whether, when and to whom personal information is released [6]. To aid in the definition of privacy, marketers often define various levels of sensitivity, ranging from 'sensitive personal' to 'non-personally identifiable' [7] and define policies as to what they may or may not do with the various types of personal information. Closely related to privacy is the notion of 'nymity'. It comes in three broad categories — anonymity (having no name), pseudonymity (having an assumed name) and verinymity (having true name). The 'nymity' categories are important, in practice, as quite often marketers blur the boundaries between them.

The Web is a 'leaky' environment, where fragments of personal data could be correlated together to profile the person behind a 'nym' (depending on the nym mode, it would therefore be possible to either correlate the profile to a real person or to a pseudonym). Taking a specific scenario where a user first visits a fictitious Web site (www.portal-example.com) that has an image (banner) served by a fictitious advertising server (www.ad-server-network.com), the advertising server has the opportunity to define a cookie on that user's browser. Such a cookie is capable of being persisted (on the user's machine) and by its value or name the advertising server could recognise the same browser over and over again, thus building a profile. This profile could consist of a list of pages the user has visited (through the 'referrer' field of the HTTP protocol), when those pages were visited, the IP address (and hence the ISP, and perhaps the time zone), etc. Also, imagine that the referrer Web site has been poorly designed and the 'referrer' URI (uniform resource identifier) has user identifiable information as part of it (this could happen, for example, when an HTTP GET form is used to submit some kind of registration information to a product provider, who would then include third party content into the response page). Then the advertising server could attribute more and more details into the compiled profile until the user moves from pseudonymous to verinymous (i.e. the true identity is discovered).

To address these problems, an opt-out mechanism is usually available, where the unique identifier is exchanged for an identifier shared by all users that have opted out. This solution is not 100% guaranteed. Examples show that bugs in implementations could well lead to the tracking continuing even if the users have opted out. Also, there are short-term identifiers such as IP addresses that could well be used to track a user's short-term session even if no cookie is used. Opting out really only refers to opting out of persistent identification. With broadband Internet access, the IP addresses are less often changed leading to ever-longer tracking sessions.

Another way to address the privacy issue is the development of the P3P standard. P3P is a machine-readable language for specifying privacy policies, i.e. what information is collected and stored by a Web site. A P3P-enabled browser could deny requests (or display warnings) from Web sites that do not comply with any user-specified privacy policy.

As we have seen, on-line privacy is a subtle issue and the distributed personalised advertising system should not leak any information (at least not any information which is not already leaked, such as an IP address). The use of random identities ensures that pseudonymity is preserved.

The banners are generated at the advertiser's platform, so the broker or the publisher are unable to see the data which caused that specific advertisement to be generated. In the described implementation, the advertisement will pass through the broker, but by this stage it is in the form of an image which is likely to be unsuitable for data mining. Also, the broker should comply with a privacy policy which will not allow him to either store the images or use any intelligent tools to data mine them. Inadvertently, some information is leaked to the advertiser — the time of day when the user is browsing.

11.7 Future Trends and Summary

The on-line industry is rapidly realising that the fragmentation of the user profile and identity creates barriers for economies-of-scale in eCommerce. Several initiatives have recently been proposed to provide a solution to that problem. In one of the proposals, a single company — Microsoft — proposes to act as a central Internet 'government', and issue 'passports' to its customers [2]. Any site can register with Microsoft to use the passport service to authenticate the users. In effect, using passports creates a single sign-on capability across the network of participating sites.

The user's profile is held centrally and the user is in control of what information is permitted to be shared, and with which sites.

Apart from the obvious disadvantage of this proposal (i.e. the single point of failure or breach), it does not directly address the problem of defragmenting the distributed user profile — i.e. many companies have already built user profiles and want to continue using them. It is unlikely that they would pass on, or even share, these profiles with a third party.

Another proposal to solve the identity fragmentation problem is found in the creation of the Liberty Alliance project [8]. It believes that federated network identity is the key to realising new business opportunities, coupled with new economies of scale. A central component of the Liberty Alliance project is a single sign-on capability [9]. This is achieved by a federation protocol similar to the 'phase one' protocol used in section 11.4. One of the advantages to having a standard proposal in this area is to ensure interoperability between vendors as well as to help gain an industry momentum.

The Liberty Alliance project should help to increase the acceptance of federated network identity/profile concepts across the industry, which in turn will assist with establishing the distributed profile advertising system as a major advertising/CRM channel.

References

1 '*FTC investigates DoubleClick's data-collection practices*' — http://news.com.com/2100-1023-237007.html?legacy=cnet

2 Microsoft Passport — http://www.passport.net/

3 Amazon's Associate Programme — http://www.amazon.com/associates/

4 Site Directory — http://www.doubleclick.com/us/product/online/sitedirectory/

5 MSN Profiler — http://advantage.msn.com/home/home.asp

6 Curtin, M.: '*Developing Trust: Online Privacy and Security*', Apress (2001).

7 Network Advertising Initiative — http://www.networkadvertising.org/

8 Liberty Alliance — http://www.projectliberty.org/

9 Liberty Architecture Overview, Version 1.0 (July 2002).

12

PERSONALISATION AND WEB COMMUNITIES

S Case, M Thint, T Ohtani and S Hare

12.1 Introduction

Web communities primarily exist so that people with common interests can communicate and exchange information with each other. Examples of such communities are:

- members of the same company;
- people sharing a common profession, e.g. trade associations;
- any group of people working on some shared enterprise.

Such communities typically have deep reservoirs of knowledge and membership allows users to tap into these. This knowledge can be stored either explicitly in the form of documents or implicitly within the minds of the members. Web communities should therefore aim to access both types of knowledge source so that members can gain the maximum benefit from the community.

In a traditional community, members control the information flow either by one-to-one communication or by broadcasting to a group. For example, certain members are responsible for informing appropriate people about upcoming events. Other members may be regarded as experts on certain topics and can answer queries or direct members to focused documents or people. Intelligent software agents offer an ideal technology for providing novel and automated services for Web communities, combining intelligent routing of documents to people, or people to people, within a single coherent framework.

This chapter describes how a set of technologies has been integrated to produce a complete Web community solution. Initially, various components existed as independent projects in different companies. BT Exact had developed the personal agent framework (PAF), a platform for integrating agent-based personalised information management services. Fujitsu Laboratories had built the open agent middleware (OAM) that manages information resources in an open environment.

Teamware Group Ltd had constructed Pl@za (pronounced 'Plaza'), a commercial platform for hosting eCommunities, with integrated services for secure registration, messaging, event posting, discussion groups, and site customisation tool-kits. A collaborative effort resulted in the Knowledge Browser™ tool-set, an integration of PAF and OAM components as a Pl@za service. The following sections describe each of these components in more detail.

12.2 System Components

12.2.1 Personalisation

Even in Web communities where the primary aim is to share information and networks with like-minded members, personalised information management services are highly desirable. Applications that tailor services to the individual improve personal productivity — for example, by reducing search and filtering time to find relevant information. Personalised services also increase the value of the underlying eCommunity beyond a social or networking environment, i.e. the community Web site becomes an attractive, permanent 'home base' for the individual rather than a detached place to go for on-line socialising or networking.

The personalisation aspects of the Pl@za and Knowledge Browser solution consist of:

- the creation and management of interest-based user profiles;
- delivery of documents specifically relevant to each user's profile;
- information about other users with interests/expertise on specified topics;
- just-in-time notification about information relevant to the current context of a user's activity.

These features are provided by the components of the personal agent framework, a unified environment for hosting personal agents and their services. It consists of a profile manager that represents the user's interests and a suite of agents that use the profile and suggest changes to it. Each agent can operate independently of other agents, but agents can, and do, share information to augment individual capabilities [1].

12.2.1.1 PAF Profile Manager

Agent services are centred around a profile of the user which corresponds to the user's long-term interests. The profile has an explicit structure, i.e. users can inspect and alter all parts of the profile providing them with full control, and it comprises a set of interest categories arranged into a hierarchy. Individual agents may make use of the hierarchical structure or treat each interest as a separate topic. Interests can be created from scratch by the user or selected from a predefined domain hierarchy. Each interest has a number of associated attributes.

- Privacy

  Each interest has one of the following privacy settings — private, restricted or public. In the case that the interest is private, only the user and their agents know about this interest. Restricted interests can be shared with a specified group of users. Public interests can be seen by any other person or agent.

- Expertise

  Each interest has a rating corresponding to the expertise of the user in the given area.

- Importance

  This attribute shows how important this interest is to the user at the current time. As the user's focus changes this can be reflected in the profile.

- Positive key-phrases

  This is a set of words and phrases corresponding to this interest.

- Negative key-phrases

  This is a set of words and phrases corresponding to exceptions to the positive key-phrases.

These attributes are all fully under the control of the user. While individual agents suggest alterations to the profile, they do not make the changes without the user's express consent.

12.2.1.2 Bugle

The Bugle agent is designed for the delivery of relevant text information to the user. Although applicable to documents in general, Bugle focuses on generating personal newspapers. It utilises the PAF profile to filter incoming news (fed from a content provider such as Reuters), and creates a customised newspaper based on the user's interests. The main page contains article summaries grouped according to the user's interest categories, with hyperlinks to the full articles. At the end of each full article the Bugle extracts characteristic keyphrases and offers them as suggestions to update the user's profile for that topic. The user can elect to add them and update the profile in a semi-automatic manner. There is also a link to the iVine agent to facilitate the search for other users with interests in that current article.

Other 'personal' news services, generally available on the Internet, push news about broader topics (e.g. sports, finance, entertainment, health). The Bugle, however, uses full-text matching between the key-phrases in the user profile and the individual news articles, thus providing a more focused personal service. It also considers the 'importance' level of various interest topics to the user, in proportioning the content of articles presented in the personal newspaper.

12.2.1.3 iVine

iVine is an agent designed to locate people with common interests to facilitate networking among community members. The user can query the iVine agent about people with interests similar to their own or people whose interests are related to specific query terms. iVine locates all the people who have relevant interests (subject to privacy restrictions, i.e. users who marked matching interests as private or restricted are not 'found'). When presenting the list of relevant users, iVine displays the search results with corresponding strength of match, expertise, importance and interest information, along with links to contact details and an e-mail editor. The extra information allows the user to conveniently verify that the suggested contact is in fact a good one.

12.2.1.4 Radar

Whereas some agents work in batch/off-line mode such as Bugle, Radar is a real-time information-finding and delivery agent. It makes use of the information sources available to the system such as news reports, documents and FAQs, and other agents (e.g. iVine) to present just-in-time information relevant to the user's current context.

Radar monitors the user's activity on a desktop application (e.g. Microsoft Word) and selects text corresponding to the user's current focused activity (such as the paragraph associated with the cursor). Radar locates relevant information and people from its sources based on the selected text. Links to the resources are then presented in a result window along with the scores, resource names, dates and titles of the results. Users can therefore concentrate on their primary work — e.g. writing a document — while Radar does the time-consuming work of searching related information in the background.

12.2.2 Resource Management

The PAF functionality is closely tied to the information sources, i.e. upon replacement or addition of new resources, parts of its code need to be changed. Thus, it is desirable to incorporate a middle-layer platform for constructing flexible, dynamic, and scalable distributed systems as offered by OAM. OAM supports collaborations of agents in a seamless and transparent way, and it enables agents and resources to be connected in a plug-and-play manner over a network. When a new agent is plugged into the network, OAM arranges the organisation of agents, adapts them to each other, and manages their collaboration. OAM enables dynamic behaviour by providing three main functions — distributed mediation, communication via field, and reflective adaptation.

In a dynamic environment, services are constantly becoming available or being removed. OAM provides multiple mediator agents to find and access the most suitable information resource at any given time [2, 3]. Mediators contain tables of condition and destination pairs for determining the suitable destination for a search request. When a service provider plugs a service agent into OAM, the meta-information about the service is advertised to a mediator in the same OAM platform. The mediator stores this information within its own table and then forwards it to neighbouring mediators.

A common communications medium should be sufficiently flexible, dynamic, and scalable to enable open communication for collaboration of agents over the Internet or intranet. The medium named field, enables such communication among the organisation of agents. As a communications medium, agents communicate with others in the peer-to-peer or multicast manner on the field. As an information-sharing medium, agents can share information between themselves on the field as a message blackboard. Thus the field works as a logical network and at the same time as a shared memory. Communication via the field is an event-driven, multicast-based communication. All agents on the field listen to all messages in the field, and each of them autonomously reacts to messages according to its own criteria named patterns. An agent becomes aware of all events and reaction to them is up to the agent. Agents on the field can be added to and deleted at any time, independently of other agents. The patterns of agents can also be dynamically changed. Thus, collaboration of the field is flexible and dynamic.

An organisation of agents must agree on their data formats and protocols before collaboration, because each agent has its own general data format and protocol. In order to unify them easily, OAM enables updating of interfaces and protocols for new interactions and collaborations in run time. In summary, reflective adaptation consists of changing the Java servlet code dynamically, which can be used to add new features to the interface, or even update existing code, e.g. by adding an extra parameter to a method call.

12.2.3 eCommunity platform

PAF and OAM provide the platforms for personalised agent-based services and distributed resource management in a flexible and scalable fashion, but additional features are needed to provide a complete eCommunity solution. Those features are found in Pl@za, a modular platform for creating interactive Web sites [4].

Teamware Pl@za is a solution for building Web sites to support communication and collaboration in intranet, extranet and Internet communities. It provides an extensive on-line tool-set for constructing a Web site to meet the needs of a community. In the case of a business for example, it can host the standard Internet presence, while also providing an interactive intranet for communication within the business. At the same time it can host an interactive extranet for enhancing communication with key customers and partners.

This framework provides built-in support for user registration and authentication, session management, access level control, on-line chats, interactive business cards™, discussion groups, special interest groups, event management, messaging, billing, and user directory/search. In addition, Pl@za also provides support for the community designers, in terms of customisable user interfaces with dynamic content, templates, form pages, and context-sensitive help for the users. These tools and services facilitate the creation of interactive Web sites tailored according to the preferences and access rights of individual users and groups.

Pl@za is built around the structure of an organisation with different types of membership and committee structures, special interest groups and team working. It is possible to define tailor-made group types that match the sets, roles and positions each user has. Each group has a manager or managers who can introduce users to the group and change the content of the group's 'home pages'. Altering or adding new content can be done easily through a standard browser and only requires knowledge of basic HTML. Content is thus kept within the full control of those who want it published.

In addition to these features, the Pl@za document folder also provides support to enable the sharing of documents between group members. Access to these folders depends on the access rights of the individual user. Members can also see the version history of a particular document. This can be combined with the notification service to alert the user when a document has changed. The Pl@za survey management function allows surveys to be managed and included in group pages and can be used, for example, in collecting customer feedback, assessing results of a training course or assessing member satisfaction. The Pl@za event management service allows users to organise and publish events on the Web. This can include both on-line and off-line events.

12.3 System Integration

12.3.1 Development of IDIoMS

A complete solution for Web communities was created through a systematic integration of the components mentioned above. The first step involved the integration of the PAF and OAM into a technique called the Intelligent Distributed

Information Management System (IDIoMS). OAM is used to manage the information resources while the PAF personal agents are used to perform the personalised information management functions. The PAF service components, i.e. profile manager, Bugle, iVine, and Radar, are connected to the communication field of OAM and communicate with each other and with the information resources via HTTP. Information sources are wrapped by OAM agents and are mapped to information requests from the PAF agents by the mediator agent. Details of the architecture can be seen in Fig 12.1.

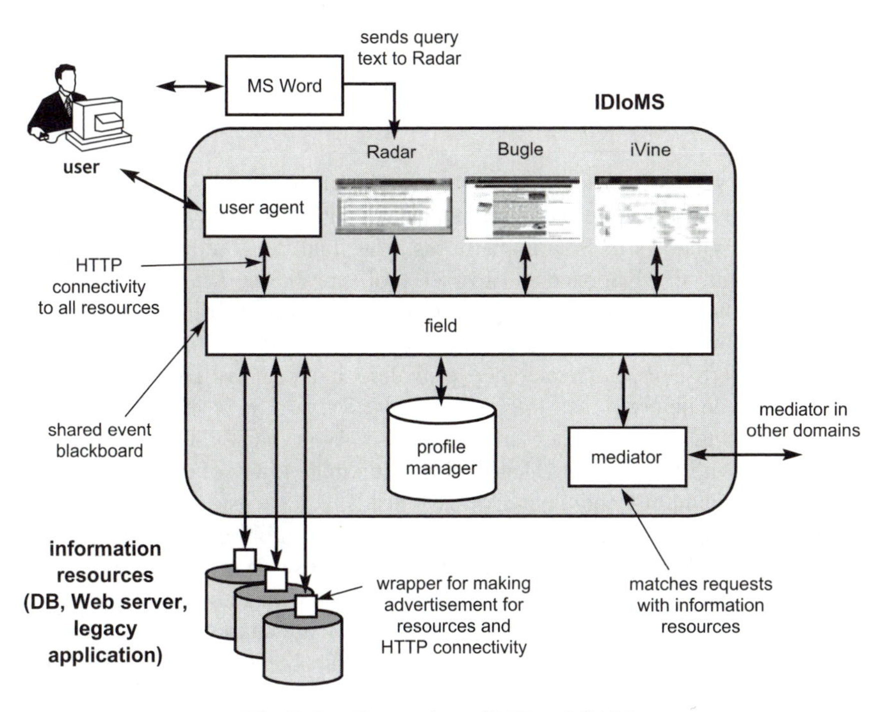

Fig 12.1 Integration of PAF and OAM.

IDIoMS was developed in Java using the Teamware Group Ltd Phoenix engine. It underwent extensive trials in Fujitsu, BT and Concert, details of which can be found in Ohtani et al [5].

Integration of these systems brings benefits to users, content providers, and service providers. From the users' point of view IDIoMS offers unified access to relevant and useful information. No knowledge is required from the user about what information sources are available and where they are. IDIoMS uses the profile and current context to locate relevant information for them. The content provider need not relocate or create special repositories for IDIoMS — only exchange data via an HTTP port. The service providers can easily add or remove a resource without otherwise affecting the performance of the services. They need only create a simple wrapper agent to convert IDIoMS requests into the content provider's internal processes. Once an appropriate advertisement (about a new resource) is sent to the mediator, IDIoMS services will start using the information resource. If the resource subsequently becomes unusable the mediator will remove the resource from its cache.

Within IDIoMS, the personal agents submit service requests to mediators who then forward them to the appropriate sources. This is a clear advantage to the service provider since each installation only requires alteration of the wrapper agents — other parts of the code-base remain the same.

12.3.2 Commercial Exploitation

Following the successful integration, testing, and trials of IDIoMS, that system itself was integrated with the latest version of Pl@za built on open Java technology and operating on Windows or Solaris platforms. The final integration effort produced Pl@za 3.5 with the Knowledge Browser tool-set. Pl@za Knowledge Browser incorporates the agent-based intelligent information services [6] developed in the IDIoMS platform as a set of Pl@za services. These services include the profile manager, expertise locator, reference provider, and personalised newsletter, as further described below.

The resulting product has an interactive Web capability that provides communities with the ability to share information in a variety of means, tailored to suit both the community and the individual's needs.

While most applications help the user find existing information in pre-specified locations, Pl@za Knowledge Browser also assists the user in reaching beyond those information sources — to grasp knowledge residing with individuals sharing the same interests and contained in other resources which are unknown to the user. This newly gained knowledge can then be propagated and shared by even more people. By connecting people with other people as well as document resources, the components of Pl@za Knowledge Browser help to build and extend the knowledge base within the Web community.

Access to Knowledge Browser services is through a standard browser, and different access levels can be assigned to Pl@za group members by the administrator.

Users see the services as an extra set of links in the Pl@za left-hand-side menu bar shown in the following series of screen shots (Figs 12.2-12.5).

12.3.2.1 Profile Manager

This functionality is accessed via the 'Manage Profile' link within the Pl@za framework, with all of its original PAF features preserved (see Fig 12.2). The user can create and edit interest topics, including associated attributes for positive and negative key phrases, privacy, importance, and expertise levels. This explicit profile structure promotes confidence and understanding for users about why they are receiving certain information (or not), and provides control on what information is being shared with community members.

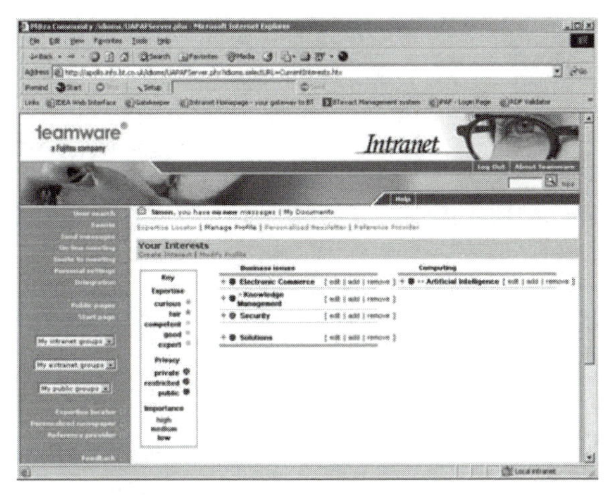

Fig 12.2 Profile management screen in Pl@za.

12.3.2.2 Expertise Locator

Finding an expert on a particular topic can save a member much valuable time in completing a certain task or locating specialised information. Even finding other non-experts with similar interests promotes healthy interaction and fusion of ideas. The expertise locator service enables community members to find others with knowledge or interest on specified topics. Upon finding relevant contacts they can readily access the users' interactive business cards or contact them through the Pl@za message centre (see Fig 12.3).

12.3.2.3 Personalised Newsletter

Latest news is of interest to nearly everyone, especially when it is about a topic of personal interest. In viewing traditional news media, one has to wait for a passage of interest or sift through a collection of articles to find the appropriate ones. In the Pl@za environment, the administrator for the eCommunity site can select content providers that are relevant to the community, while the personalised newsletter (see Fig 12.4) further selects specific articles for individual members. After reading an interesting article, users can locate other members who may be interested in similar topics and seek help or initiate further discussions.

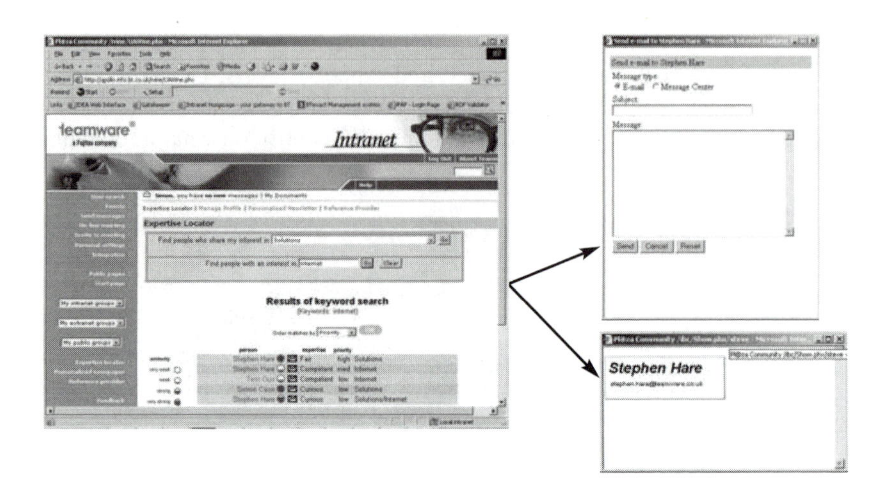

Fig 12.3 Expertise locator service in Pl@za.

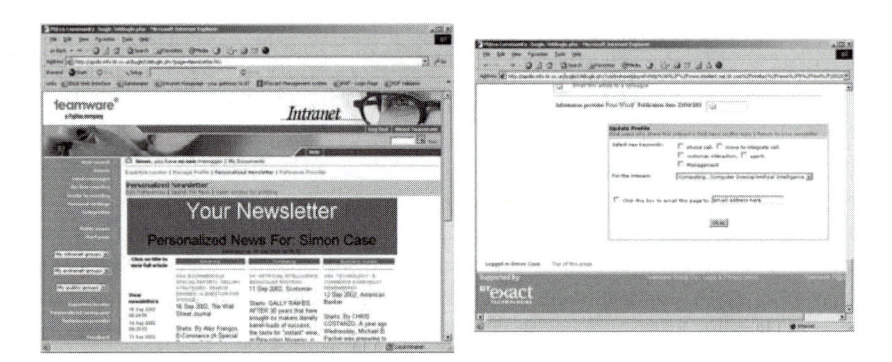

Fig 12.4 Personal newsletter screen shots.

12.3.2.4 Reference Provider

It is very helpful to have a personal assistant who monitors activities 'over the shoulder' and informs you about resources (documents and people) that are relevant to your current activity.

The reference provider service (see Fig 12.5) provides such assistance while a member authors a document in Word — the agent locates documents and contacts from the Web community and informs the author via displays in a background window.

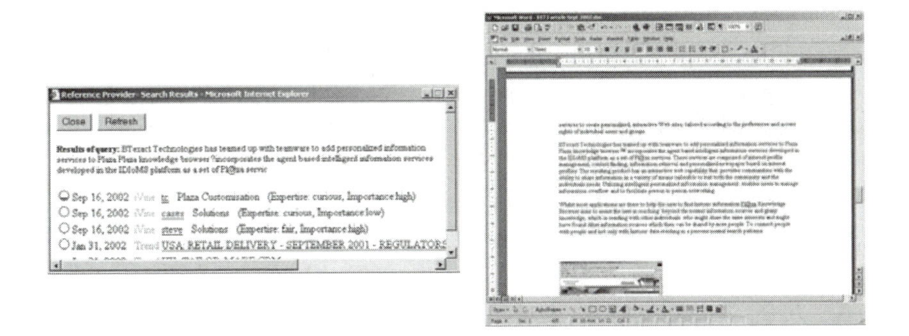

Fig 12.5 Reference provider screen shots.

12.4 Summary

This chapter has described a software system that offers a complete solution for Web communities, in terms of an integrated package for communal as well as personalised services. The Pl@za 3.5 with Knowledge Browser tool-kit comprises components for discussion groups, on-line chats, shared folders, event management, surveys and messaging, as well as personalised services (profile manager, newspaper, contact finder, reference provider).

The personal agents help to promote site 'stickiness' and brand loyalty, while the Pl@za Knowledge Browser is an example of a successful integration of personal agent technology, intelligent middleware, and flexible eCommunity platform into a commercial product.

The integration was conducted in a manner that preserved the strong features of the independent sub-systems, while the final system operates as one seamless unit resulting from the co-operation between all modules.

References

1 Crabtree, I. B., Soltysiak, S. J. and Thint, M.: '*Adaptive personal agents*', Personal Technologies Journal, **2**(3), pp 141-151 (1998).

2 Mohri, T. and Takada, Y.: '*Virtual integration of distributed database by multiple agents*', in Proc of the First International Conference, DS'98, Lecture Notes in Artificial Intelligence, **1532**, pp 413-414 (1998).

3 Takada, Y., Mohri, T. and Fujii, H.: '*Multi-agent system for virtually integrated distributed databases*', Fujitsu Scientific and Technical Journal, **34**(2), pp 245-255 (1998).

4 Case, S., Azarmi, N., Thint, M., and Ohtani, T.: '*Enhancing eCommunities with agent-based systems*', IEEE Computer, **34**(7), pp 64-69 (2001).

5 Ohtani, T., Case, S., Azarmi, N. and Thint, M.: '*An intelligent system for managing and utilizing information resources over the Internet*', International Journal of Artificial Intelligence Tools, **11**(1), pp 117-138 (March 2002).

6 Soltysiak, S., Ohtani, T., Thint, M. and Takada, Y.: '*An agent-based intelligent distributed information management system for Internet resources*', INET (2000).

13

LOCATION INFORMATION FROM THE CELLULAR NETWORK — AN OVERVIEW

W Millar

13.1 Introduction

The deployment of mobile networks, coupled with the rapid development in handsets and other mobile computing devices, has enabled a rapid growth in the ability to deliver advanced services to the user on the network periphery. Mobile networks are no longer considered to be a transport mechanism exclusively for voice calls — growth in data services has been significant. This growth has come initially from SMS traffic, but is increasingly represented by traffic from more advanced applications such as Internet browsing and e-mail access. Currently these applications extend the reach of the desktop and provide additional means of accessing traditional information. However, there is now a trend to see new applications which are enabled by the deployment of new technology, and which are delivered to a receptive audience which is becoming increasingly familiar with the underlying technology.

Two generic areas of service provision are emerging — location-based services in which the location of the mobile user is used as a parameter for service provision, and personalised services in which a personal profile can also be used as a parameter governing the supply of information to the user (see Chapter 1 for more details on personalisation). The profile could also be extended to include aspects of the network (e.g. available bandwidth or delay) and of the device (e.g. local memory, screen size and colour). This chapter examines how location-based services can be delivered with current technology, and examines two proof-of-concept demonstrator systems which are showing how this technology can be used. The chapter begins with a short general overview of location-based technologies for reference purposes, and then focuses on the use of the mobile location computer (MLC) from O_2 as a means of obtaining location information from a mobile handset. The information returned from the MLC as a result of a query is described, together with a test environment used to compare the output from the MLC with more accurate positional information obtained from a GPS. The chapter concludes with a description of two proposed applications to make use of the service.

13.2 Overview of Location-Based Technology

Perhaps the best known example of positioning technology is the global positioning system (GPS) developed by the US military for defence purposes. This system is based on 24 satellites which orbit the Earth, thereby enabling 4 of them to be viewed from any point on the Earth's surface at any time.

In operation, users wishing to know their position must have a GPS receiver capable of receiving signals from the satellites. If the user can receive signals from at least three satellites, then, from the measurement of the distance between the user and the satellite, the exact location of the user on the surface of the planet can be found. The most difficult part is determining the distance of the user from each satellite. This is measured indirectly from the time taken for the signal to reach the user from each satellite, coupled with a knowledge of the speed of propagation, which is then used to calculate the distance. The GPS receiver is capable of making these calculations and thence determining the position in terms of latitude and longitude.

In standard terrestrial wireless communications networks, the mobile handset is portable and does not have to be at a fixed location in order to work. The location of each handset relative to the network is maintained within the network so that calls can be routed to it. However, designers, when planning the networks which are currently in use, did not consider making positional information available outside the network, probably because it was perceived as not adding any value and hence the complexity in installation of any suitable technology was not justifiable. Thus for current networks, any capability to make location information about the handset available at the network periphery for use by service providers must be in the form of an addition to the existing infrastructure. The provision of this additional functionality is subject to the normal business case processes and in to-day's commercial environment, the capital cost of the provision of the required equipment is a significant factor in the successful outcome of the business case. Thus, as will be discussed later, while technologies exist which can provide relatively accurate positional information about a mobile handset to the network periphery, in an uncertain commercial environment these are unlikely to be deployed as part of the current network infrastructure. One external influence which may change this is the introduction of legislation. In the USA, the FCC has stated that by the end of December 2005, 95% of all handsets sold must be location compatible to allow users to be found in an emergency, such as a car accident. This is required to be to an accuracy of 50 m for 67% of calls and 150 m for 95% of calls. European legislation is in progress, although the consideration of the location-based element has been deferred to 2005 for a number of technical, regulatory and business reasons.

Technologies which can be used to augment the information available from mobile devices can be categorised into network-based methods and handset-based methods. In the former category are the following.

- Time difference of arrival

  This uses accurate clocks to determine the difference in time by which radio signals from the handset reach different cell sites. This time difference can then be used to determine position, velocity and heading. This technique uses existing cell towers and infrastructure but requires specialised receivers to be placed at the base-station.

- Angle of arrival

  This measures the direction of a signal received at multiple cell towers with respect to the antenna. Since the positions of the antenna are known, the position of the handset can be calculated relative to them. Again this requires the installation of directional antennas on the cell towers.

- Multipath analysis

  This makes use of a predefined multipath database for a specific service area. The location of the handset can be determined by comparison between the data received from the handset and that already measured and stored in the database. This approach also requires the installation of specialised receivers in the base-station and the construction of the multipath database from measurements.

Handset methods include the following.

- Global positioning system

  In this mode the receiver is augmented by the inclusion of a GPS receiver and antenna on the handset. The GPS receiver must be able to see the minimum number of satellites to produce a positional fix and it may not be immediately obvious to a user that this is not the case.

- Advanced forward link trilateration

  This is a time difference of arrival technique which requires handsets with precise timing capabilities. Measurements are sent back to a location processor in the network which calculates the location of the handset.

- Enhanced observed time difference

  This is similar in approach to the GPS system, except that the timing signals are sent from cellular base-stations and software in the telephone calculates the position.

13.3 The Mobile Location Computer

The mobile location computer (MLC) has been introduced by O_2 as a way of making the cell of origin information available to the periphery of the network. The MLC is based on Redknee [1] technology and, as a result of a query made to the MLC, returns the currently known location of the specified handset or an error message. The position is returned in the form of latitude and longitude co-ordinates derived from the cell-location information and an associated estimated radial measurement within which the handset is expected to be located. The format of the message from the MLC is described in the published developers' guide [2].

Connection is made to the MLC infrastructure through the use of a virtual private network (VPN) connection. This ensures that only registered users of the service can attach to the MLC. Before attachment can be made, a sequence of tests must be undertaken to ensure that the application software will not overload the MLC and that appropriate mechanisms exist to throttle back the traffic to the MLC and hence prevent traffic overload.

In addition, it is important to establish that error conditions are trapped and dealt with in an appropriate manner. BT connects to the MLC using existing software as a gateway both to access the UK operators' SMS networks and to provide most of the required functionality which will allow an approved connection to made.

At the time of writing, all O_2 network operators have an MLC installed in their networks, although it is not possible to get location information from a telephone roaming on to the network. Additional information on the MLC and its associated interface can be found on the appropriate Web site [3].

13.4 MLC Accuracy

One major question about the MLC is, of course, the accuracy of the information returned to the user. The location information returned is a position derived from the location of the cell base-station with the radial estimate of proximity to the absolute position.

It is known from the outset that the precision of the positional information returned from the MLC is a few orders of magnitude less accurate than that obtained from the GPS system, i.e. up to a few kilometres instead of the few metres which is possible from GPS. However, an outstanding question is: 'How accurate is the MLC and consequently how useful is it?'

In order to answer this, a test environment has been established which allows simultaneous measurement of the handset position by the MLC and by a GPS fix. This information is returned to a central platform via an SMS message and the results stored for subsequent processing. An architectural diagram, showing the associated message flows of the complete test environment, is illustrated in Fig 13.1.

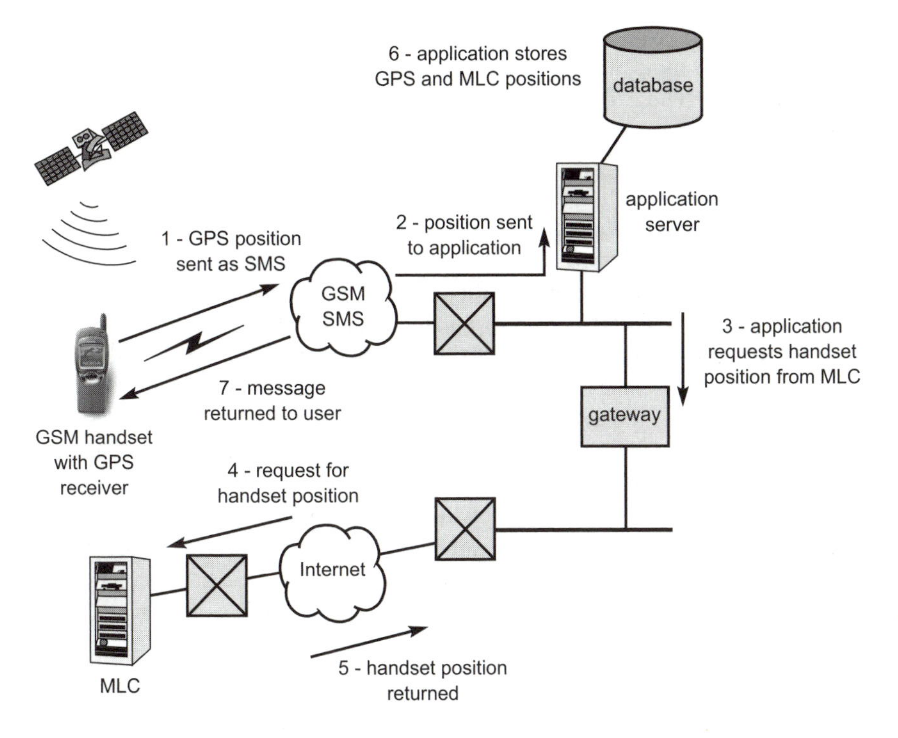

Fig 13.1 Architecture of test environment to measure MLC accuracy.

The device used to perform the measurement is a Benefon [4] telephone which has a built-in GPS receiver. This telephone can record GPS position information and make it available for onward transmission within an SMS message. In operation, the user sends a position fix as an SMS message to an application which is running on an application server attached to the BT SMS gateway. The receipt of the SMS message from the Benefon telephone triggers an event in the application to request the position of the handset from the MLC. On receipt of this request, the MLC determines the position of the handset and returns this information to the application. The application now has two positional fixes on the handset, an accurate one from the GPS network and a less accurate one from the MLC. Since the MLC position is recorded as being within an estimated distance of an absolute latitude and longitude position, then simple co-ordinate geometry can be used to determine if the GPS position measured is within the distance from the absolute latitude and longitude position obtained from the MLC. This processing is undertaken and a decision made as to whether the MLC is correct or not. Once the processing is completed, an SMS message is constructed to record both positions and return them to the user together with an indication as to whether the MLC was correct. More importantly, this information is recorded and made available for subsequent analysis.

Typical results from this test environment are shown in tabular form in Table 13.1 and an indicative sample of recorded positions are shown overlaid on a map in Fig 13.2. These data sets were obtained from positions recorded from a moving vehicle on a trip in England. All measurements were obtained from a correct GPS positional fix.

In Table 13.1, the results are tabulated as follows. The first column shows the time at which the measurements were made. The second column records the radial accuracy as reported by the MLC, i.e. the MLC estimates that the handset is within this radial distance of the absolute position. The third column calculates the distance between the highly accurate GPS position and the absolute position as recorded by the MLC. If it is accepted that the GPS signal is the true position due to its inherent accuracy, then this distance can be regarded as the true radial position of the handset from the absolute position as recorded by the MLC.

If this calculated radial distance is less than the distance estimated by the MLC, then the MLC measurements can be viewed as correct; if the measured radial distance is greater than that estimated by the MLC, then the MLC can be regarded as being in error. Whether this is significant or not is an open question, since this will be dependent on the application making use of the data. The fourth column tabulates the error as a distance value in this case, or records that the MLC is correct.

Table 13.1 MLC accuracy figures.

Timestamp	Radius reported by MLC	Calculated distance between GPS and MLC positions	Error (m)
Tue Sep 24 08:31:10	4427	1409	Correct
Tue Sep 24 08:37:55	4427	1107	Correct
Tue Sep 24 08:43:21	4427	4160	Correct
Tue Sep 24 08:46:07	Error	Error	Error
Tue Sep 24 08:49:08	7000	4553	Correct
Tue Sep 24 08:51:34	7000	7394	394
Tue Sep 24 08:53:54	7000	1319	Correct
Tue Sep 24 08:56:21	3883	4124	241
Tue Sep 24 08:57:34	3883	4436	553
Tue Sep 24 09:02:22	3883	2589	Correct
Tue Sep 24 09:04:44	3883	2776	Correct
Tue Sep 24 09:07:01	3883	1734	Correct
Tue Sep 24 09:08:15	3883	3592	Correct
Tue Sep 24 09:11:27	3883	2932	Correct
Tue Sep 24 09:18:08	3883	456	Correct
Tue Sep 24 09:19:22	3883	2937	Correct
Tue Sep 24 09:22:06	1595	3978	2383 (see text)
Tue Sep 24 09:29:50	1595	1428	Correct
Tue Sep 24 09:37:43	1595	1658	63
Tue Sep 24 09:39:49	3212	2594	Correct
Tue Sep 24 09:43:08	3212	3122	Correct

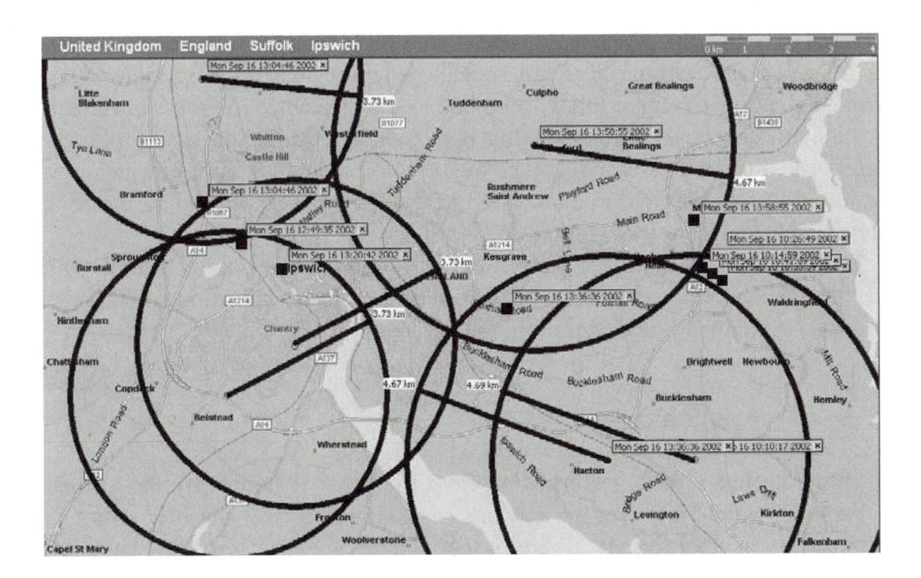

Fig 13.2 Map showing GPS position recorded (small squares) compared to MLC position recorded (large circle with radial line). The inner end of the radial line indicates the absolute position returned and the radius represents the estimated maximum distance from this absolute position.
[Map information '© 2001 Crop' prepared using Microsoft MapPoint Europe 2002]

There are a few factors which govern the accuracy of the information obtained from the MLC. The most obvious one of these is the variation in cell size. In the UK network cell sizes can vary from a diameter of a few hundred metres up to a diameter of several kilometres in rural areas. Consequently this variation in size introduces a limitation in the accuracy which can be obtained. Since no directional information can be obtained from the position of the handset relative to the cell base-station, the handset position can only be determined to be within an estimated distance from the absolute position of the base-station. While the current position of the handset is obtained by the MLC from the handset when a request is made, i.e. it is not the last known position which is returned, there are some short delays in transferring the information from the handset to the MLC and subsequently to the requesting application. These are limited to a few seconds maximum and are indeterminate in nature. This in turn means that, if the handset is stationary, then an accurate position, within the limitations of the MLC, is returned. If, however, the handset is moving, as it would be in a vehicle, then by the time the position is picked up by the application, the handset could have moved. Observed delays of several seconds can be incurred in the MLC. In a rural area where speeds are more likely to be higher than in an urban area, this movement can have the effect of generating large errors from the MLC, since the cell boundary can be crossed in that time. This type of error can be seen at row 17 in Table 13.1 (highlighted) where the time lag between recording the GPS position and the MLC position has (most likely) enabled the handset to move from cell to cell, thereby generating what is recorded as a 'large error'; however, it is more likely to be due to the difference in times at which the location measurements were made. In a deployed application, this error will be less significant, since the position recorded by the MLC reflects the time at which it was requested to be done by the application and so is more accurate than in the test environment.

These limitations in accuracy constrain or define the types of application which the information, returned from the MLC, can be used to generate. Some of these will be examined in the following section.

13.5 Applications

Since the accuracy of the information from the MLC is limited by the available technology within the cellular network, it is important to consider using this information in applications where it can either augment existing sources of information to improve the accuracy or where the location information can be a 'useful to have' feature of the application, but not a necessary one. The BT customer base is made up of major corporate and top-end SMS customers, which also provides a focus on the types of application which the MLC can be used to generate. It is important to note that raw location information is not in itself of much value, it only becomes of potential value when integrated with other information into an overall system. With these constraints in mind, some potential applications are described below.

13.5.1 Hotel Booking

This application is aimed at corporate travellers who may be on the road attending business meetings and find themselves having to stay overnight unexpectedly and thus in need of a room. Likely to be members of an executive club attached to a hotel chain, they want to find the nearest hotel to their current location and book a room. In this case the components in the application comprise an interactive voice response system to handle the user interface, the MLC to return the location information, a gazetteer to convert the latitude and longitude information into a geographic location, and the hotel booking system. Travellers can telephone their executive club reservation number, knowing that they will be connected to an automated system which will book them into the hotel nearest the current location. The application will be able to pick up this location from the MLC, send it to the gazeteer to return the geographical information, which can then be sent to the central hotel booking system. Since callers are members of the executive club, credit card details will be known together with other personalised information, such as newspaper requirements, the need for a non-smoking room, etc. For the hotel, this provides a low-cost means of receiving the booking, and, for the hotel user, it provides an automated, and hence guaranteed, mechanism for doing the booking.

It is probably fair to say that the jury is out on the value of the location information in general in a hotel booking application; certainly it is unlikely to be of much value for someone booking some time in the future from a distant location. For the scenario described above, where users have an immediate need for rooms and wish to make a quick booking, this application can provide all the necessary information. Extensions of the application would be to provide direction details on an SMS message and provide a secure access code to allow the door to be unlocked, without having to go through the hotel reception.

13.5.2 Logistics

Many large road logistics operators have their fleet fitted with onboard telematics equipment which allows them to be in contact with the vehicle and drivers at all times. These in-house vehicles can justify the cost of fitting GPS location equipment in order to provide up-to-the-minute positional information. Knowing positional information allows the centralised management personnel to make decisions about the vehicle scheduling and to take appropriate action as a result of externally or internally generated changes to the vehicle's route. This information has become increasingly useful to fleet operators since it can improve the overall efficiency of the fleet operation as well as provide an improved service to their end customers, since questions about location of deliveries and estimated time of arrival can be answered with confidence.

Some fleet operators contract other operators into their fleet in order to handle fluctuations in demand but cannot justify the cost of fitting the telematics equipment. In this scenario, the mobile handset can be used to provide location information to the management team. Of course, it would be possible to make a voice call to the driver to determine position, but increasing concern about road safety, coupled with disturbance during rest periods allows the MLC to provide an alternative and more discreet way of obtaining the location information from the vehicle.

13.5.3 Roadside Vehicle Recovery

Roadside vehicle recovery operators generally need to determine three pieces of information from their customers before being able to despatch a recovery vehicle:

- driver details and whether they are a current member of the recovery organisation;

- location;
- details of fault.

The driver details are easily determined from the membership number. The fault details are determined from a dialogue with the driver and can be categorised through a script. Location information is more difficult to determine, since the driver has usually broken down in a location where he/she did not intend to stop and consequently may not exactly know their current position. Information from the MLC will not be sufficiently accurate to resolve whether the vehicle is on the northbound or southbound carriageway of a motorway. It does, however, have the potential to assist the call centre operator in finding the location in a dialogue with the driver, particularly if the location information can be overlaid on a map of the area and landmarks determined from the map can be identified by the driver.

In this application, the objective of the recovery operator is to reduce the call-handling time while maintaining a high quality of service to a potentially highly distressed driver. This can be achieved by determining the location of a driver calling from a mobile handset during the initial dialogue with the driver and making this available to the operator during this part of the dialogue. Further questioning based around this information can then be used to determine the position of the vehicle.

13.5.4 Potential of the MLC

These scenarios describe the potential of the MLC in real-life situations and represent customer opportunities which can be identified now. The issue for this relatively new and, in some cases, niche technology for the corporate user is to identify and provide the applications which offer real and quantifiable business benefits. This is currently unknown territory and hence customers are seeking ways of trying out this technology in a relatively simple environment. For this reason BT has chosen to implement an interface to the MLC which allows simple access.

13.6 Future Directions

The scenarios described in the previous section highlight the potential for the MLC as a component in applications. The issues as already highlighted are ensuring that business case justification can be produced for the incorporation of the component into the overall solution. The acceptance process for connection to the MLC has already been discussed and the implications which it has on applications developers.

In order to enable a simpler connection to the MLC, BT has reused existing aggregator software which deals with the error-handling and flow-control aspects of the interface. Two additional interface elements have been developed, namely an http-based interface which makes use of http post and http get functions in order to send and receive messages from the MLC and a Web Services interface. In this latter interface, a SOAP interface has been produced by wrapping the existing API, and an accompanying WSDL file has also been produced to comply with the Web Services standards. In this way users can access the MLC through the BT platform as a Web Service and incorporate it into other applications. Billing for use of the service is handled through existing event logging in a billing record which is then passed to a billing engine for aggregation and incorporation into a bill. In effect, the Web Service provides a 'tell me the current location of the specified mobile handset and send me the bill' function. In the future, this is likely to be a common way of using the MLC — as a utility service within other applications.

13.7 Summary

This chapter has described some technologies which can be added to the existing 2G cellular networks in order to make location information about the mobile handset available for use in other applications. For practical and commercial reasons the only currently viable technology makes use of the cellID information and maps that to latitude and longitude information with an indication of accuracy. Work has been done to measure the accuracy of the MLC and compare it to that provided by the more expensive and more accurate GPS system. Preliminary conclusions from this work suggest that the MLC information is sufficiently accurate to be trusted, provided constraints of the measuring conditions are taken into account — such as the effect of requesting the position of a moving device. Currently the killer corporate application awaits discovery. In order to facilitate this process, BT has made available a simple interface to the MLC which allows application integration through the use of an http or a Web Services interface.

References

1 Redknee — http://www.redknee.com/products/

2 '*Location API — application developers guide*', available from O_2 — http://www.o2.co.uk/

3 Source O_2 — http://www.sourceo2.com/

4 Benefon — http://www.benefon.com/

14

A MULTI-AGENT SYSTEM TO SUPPORT LOCATION-BASED GROUP DECISION MAKING IN MOBILE TEAMS

H L Lee, M A Buckland and J W Shepherdson

14.1 Introduction

One of the main issues for the development of a group decision-support system (GDSS) for mobile teams is the mobility of the team members which makes it difficult to identify their locations and co-ordinate their input to collective decisions. Therefore, a key challenge in developing a GDSS for mobile teams is to continuously track the location of each team member and to perform appropriate actions where necessary to maintain the consistency of the decision-making process. Furthermore, mobile workers lack the information required for prompt and accurate decision making when they are not able to access the various information sources within their corporate intranet. Even when each team member is equipped with a mobile device and wireless connection, searching for the necessary information is inconvenient because of connection instability, lower processing capability, inconvenient user interfaces (e.g. virtual keyboard), etc.

In this chapter, an agent-based approach to developing location-based asynchronous group decision support systems is proposed, mPower (a system based on that approach) is introduced, and the way in which the above challenges can be resolved is discussed.

The mPower application adopts a multi-agent system (MAS) approach. The MAS is considered a key technology to support distributed teams. Intelligent, autonomous agents can collaborate to provide opportunities to reduce communications costs, combat information overload, and improve response times [1, 2]. The mPower system contains agents that interact to support distributed group decision-making processes. Each user of the system is supported by an agent that acts as their personal assistant, tracking their current position and using that information to automate part of the group decision-making process. The personal agent behaves independently of the user, according to a predefined decision-support policy, negotiating with other agents to exchange information as and when necessary. The autonomous characteristics of multi-agent systems are essential to facilitate asynchronous GDSS, as occasionally users cannot contribute in a timely manner to all the decision-making processes in which they are involved.

GDSSs must offer flexible support for group decision-making processes as each process has different information needs. The proposed approach to providing such flexibility is to use predefined service building blocks (referred to as generic service components) which use FIPA [3] standard interaction protocols to facilitate inter-component communication. The generic service component for multi-agent systems (GSCmas) is an aggregation of role-based [4] software components, with each GSCmas supporting a different group decision-making process. Therefore when a user needs to participate in a novel group decision-making process, the user's personal agent can support that process via the appropriate GSCmas.

This chapter is structured as follows. The next section reviews related literature. The third section details the architectural features of GSCmas that particularly lend themselves to use in asynchronous group decision-support systems. In the fourth section, there is a description of the use of mPower by a mobile team in a telecommunications company in the UK. The chapter concludes with a discussion and summary.

14.2 Literature Review

Multi-agent systems are used as a core technology in various applications, from information retrieval [1] to business process automation [2]. Many of the multi-agent system platforms are based on Java and can be run on heavyweight devices using Java 2 standard edition (J2SE) [5, 6]. However, the lightweight extensible agent platform (LEAP) is an exception, as it enables the key components of a multi-agent system to run on a wide range of devices, from PDAs and Smart Phones using either Java 2 Micro Edition (J2ME) or Personal Java, to desktops and servers running J2SE [7]. This enables the benefits of agent technology (e.g. autonomous decision making based upon contextual information) to be applied to mobile applications including location-based services.

LEAP, used as the basis for mPower, is a multi-agent platform that both complies with recognised standards for agent systems, i.e. FIPA, and can run on lightweight devices. In addition, LEAP provides useful functionality for mobile communications, such as the mediator concept and the JICP protocol [8].

Group decision-support systems fall into the category of CSCW (computer supported co-operative work). CSCW '... is about groups of users — how to design systems to support their work as a group and how to understand the effect of technology on their work patterns ...' [10]. GDSS can be classified as asynchronous or synchronous, according to the way in which collaboration takes place. A synchronous GDSS typically provides an electronic meeting place where participants can collaborate in real time, complemented by some computerised facilities that provide relevant information to participants during the course of the decision-making process.

An asynchronous GDSS generally provides an electronic discussion space where participants can read or write items relating to a particular issue at different times, and therefore do not necessarily participate concurrently. A summary of the various types of GDSS can be found in Khoshafian and Buckiewicz [9].

The mPower system is an asynchronous GDSS in that the mobile workers participate in a decision-making process at different times. However, it does not provide a shared place where the participants can read or write items. The views of the participants are passed to each other via messages exchanged by autonomous agents.

Current GDSSs are mostly based on wired environments where the participants are guaranteed to have a stable connection and access to various information sources with reasonable response times. However, this assumption does not hold for mobile workers, who frequently have difficulties getting information from multiple information sources because of wireless network instability. This chapter shows that many of the challenges faced by designers of asynchronous group decision-support systems for mobile working environments can be addressed by the combination of multi-agent systems and GSCmas.

14.3 mPower — an Asynchronous GDSS for Mobile Teams

14.3.1 Overall Architecture

The main components of a decision-support system (DSS) are data, dialogue, and model (the DDM metaphor) [10]. The model component represents the basic decision model used in the system. The data component is used during the execution of the decision model, and the dialogue component is used to get input from, or show model execution results to, the user. The mPower system uses the DDM metaphor within a multi-agent framework.

A GDSS is different from a DSS in that a GDSS supports multiple users having multiple roles, whereas a DSS supports a single user with a single role. Within a GDSS the different roles have associated activities that are linked to form decision processes.

Figure 14.1 shows the internal architecture of an mPower personal agent that supports group decision-making processes. A personal agent consists of four main components:

- dialogue management module;
- decision model executor;
- data management module;
- GSCmas library.

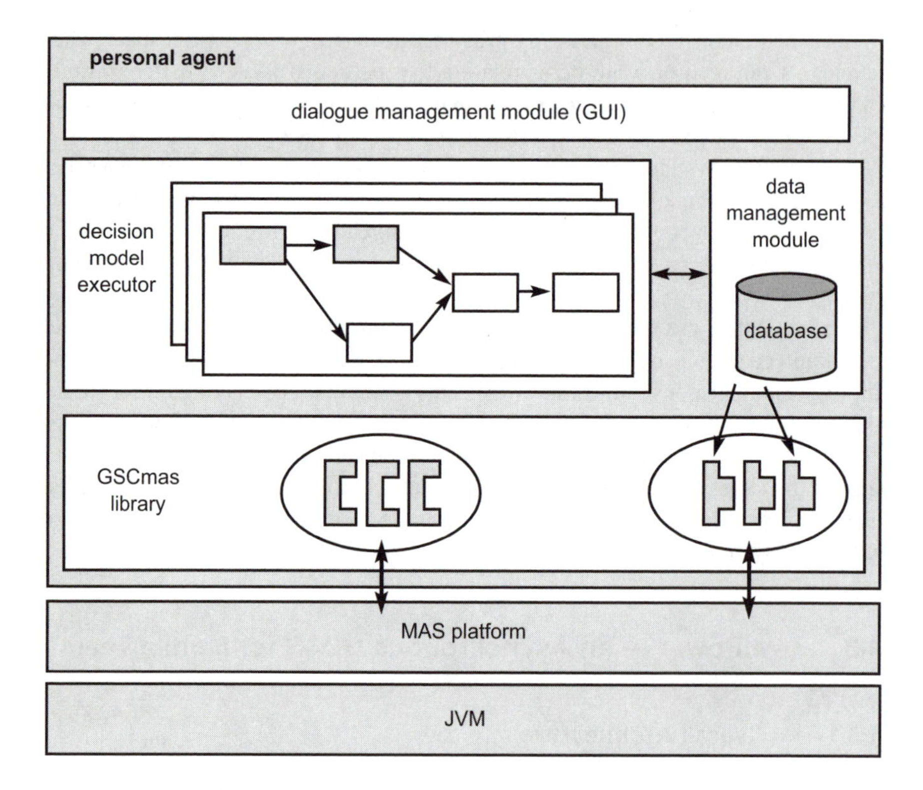

Fig 14.1 Internal architecture of an mPower personal agent.

The dialogue management module is used to provide a means to get information from, or show results to the user. The decision model executor uses the appropriate template to perform a given decision process — a decision process template specifies all the activities associated with the role that a personal agent is playing in a particular group decision-making process. During the execution of some of these activities, the decision model executor may need external input. This is obtained from the user via the dialogue management module or from other information agents. In the latter case, the data management module first uses the GSCmas library to find the generic service component type that can supply the required information, then looks up the address of an instance, and finally contacts the instance to request the information. All information obtained from external sources is stored in a database within the data management module. The fourth component, the GSCmas library, is responsible for the registration and de-registration of all GSCmas modules associated with the platform.

14.3.2 Group Decision-Making Process Definition

A group decision-making process is modelled via extended Petri nets [11], using the constructs shown in Fig 14.2.

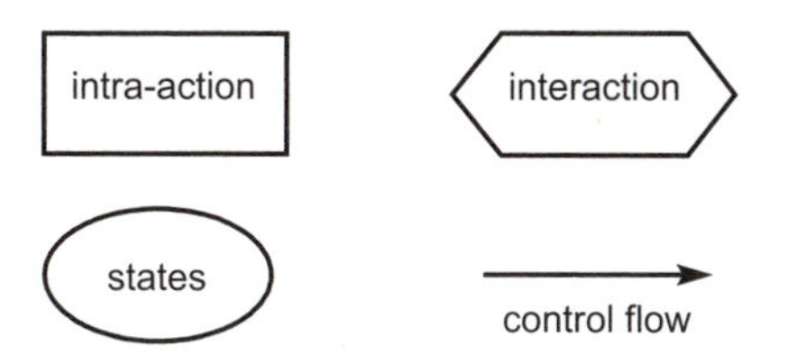

Fig 14.2 Modelling constructs for group decision-making processes.

A decision process definition is composed of nodes and links. Nodes represent activities and states of a decision process, and links represent control flows between two nodes. An activity is classified as an intra-action or an interaction. An intra-action represents an activity that can be performed by a personal agent using its own resources (e.g. evaluating a proposal based upon a policy). An interaction is an activity which requires communication with other agents, for example sending a message to, or receiving a message from, an information agent. An interaction may be specified with a GSCmas. States represent the current state of a decision process, and are used to determine the next activity that should be performed after an activity has been completed.

14.3.3 GSCmas — a Component-based Information Retrieval Process for Multi-Agent Systems

A GSCmas is a set of role-based components which interact with each other via messages to achieve a common goal (i.e. support a given group decision-making process). It is an implementation of an interaction pattern [12] for a multi-agent system. While an interaction protocol defined by FIPA specifies the order of valid messages for an interaction, the interaction pattern in this chapter specialises the interaction protocol by specifying actions that should be performed by participating agents in an interaction. Each sub-component of the GSCmas is responsible for the behaviour of a given role in the interaction and is distributed across one or more agents according to the application requirements. This means that an agent can participate in any predefined interaction by installing and using the necessary sub-component. The GSCmas can be reused in similar applications where the same interaction protocol and the same message content language are used, with minimal change to the existing software components. A GSCmas is non-divisible in that a sub-component cannot achieve the designated goal without interaction with other sub-components.

Figure 14.3 shows the overall structure of a GSCmas. It consists of an instantiation of each of the two generic role components — initiator and respondent. The initiator component is responsible for starting an interaction with other agents which have the relevant respondent component installed, and returning the service result (if any) to the requesting agent. The respondent component is responsible for handling a request message from the initiator component, and interacts with its host agent via a predefined interface to generate service results which it eventually returns to the initiator.

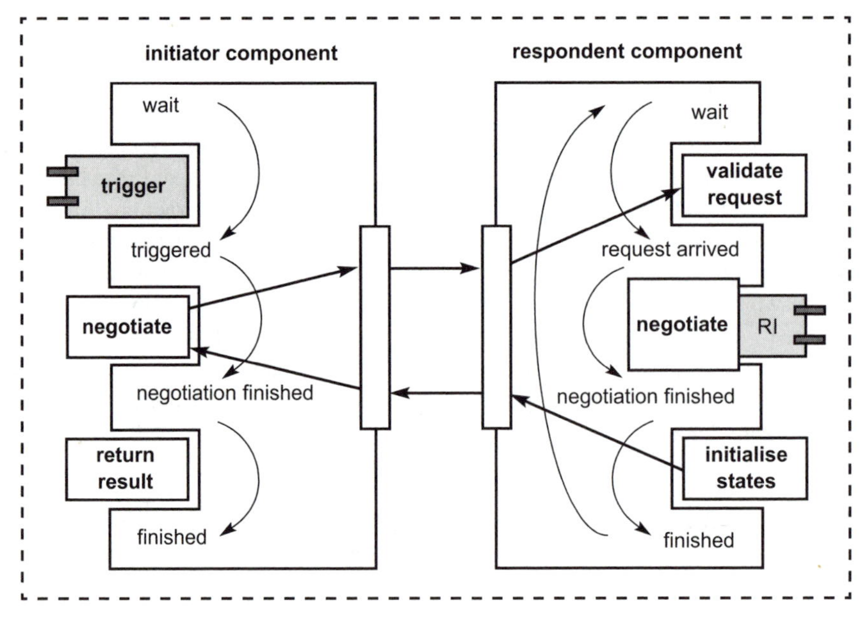

Fig 14.3 Structure of a GSCmas.

The initiator component provides a pseudo method-call interface (shown as 'trigger' in Fig 14.3). The requesting agent calls the trigger method, and the initiator component blocks (or freezes) the caller until the service result is prepared and returned. This is to align the synchronous method call of the requesting agent with the internal asynchronous message passing between the initiator and respondent components. The initiator component hides all the interaction details (such as preparing request messages, interpreting response messages, and the interaction protocol used for the collaboration with other agents, etc) from the caller. Customisation of a respondent component for a particular service is performed by providing a different implementation of its interface specification — thus different sources of the same information can be made available to initiator components.

Aggregation produces more complex service components which use service results from two or more GSCmas in a certain order. Figure 14.4 shows a GSCmas which uses the service results from two generic service components ($GSCmas_2$ and $GSCmas_3$). The links in this GSCmas 'chain' interact with each other via the trigger methods provided by each initiator component.

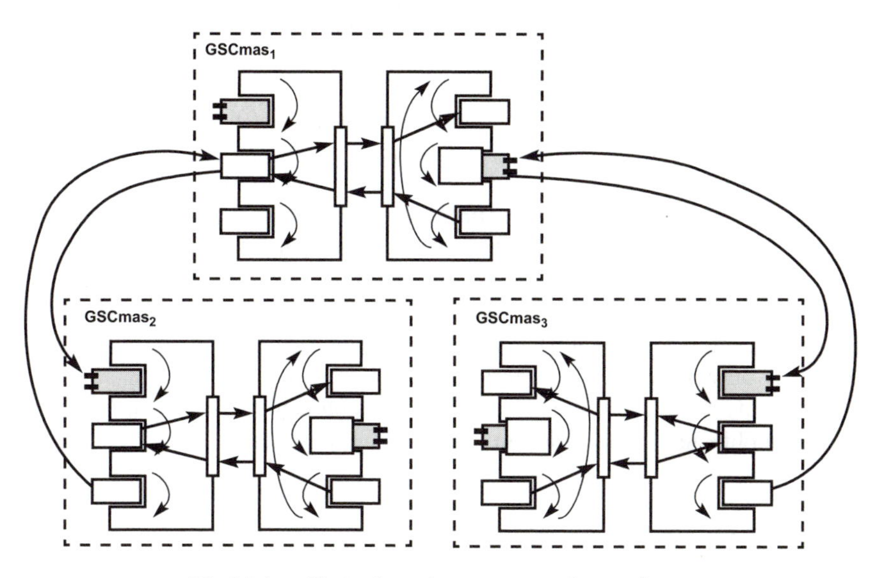

Fig 14.4 Chained service component interaction.

14.4 Illustrative Example

14.4.1 Target Application

The mPower application has undergone trials with a team of mobile workers in a telecommunications company in the UK. The target process was a 'survey process for telecommunications service provision' where the main actors are survey officers.

The function of a survey officer is to survey a customer's site to gather the information needed for the provision of telecommunications equipment. Typically surveys are initiated as a result of a request for service provision from a customer where there was inadequate data about the site, plant or capacity. Survey officers work within geographical areas delineated by groups of telephone exchanges, known as a ‘patch’, which vary in physical size. Survey officers each have ownership of a queue of work for the patch for which they are responsible. Jobs are allocated to the survey officer’s queue by a system that is responsible for breaking down a service provision into a number of co-ordinated tasks. New surveys are regularly allocated to the queues as customer orders demand. The number of jobs within a queue at any one time is dependent upon the service provision demand at the time. Factors that influence this include marketing of new communications services and patch population.

One of the main decision problems for survey officers is dynamic job allocation to team members. Survey officers choose a number of jobs in the morning from their job queue, depending on the importance and geographical adjacency of the jobs. Problems can occur when urgent jobs arrive during the working day. Normally, any job (urgent or otherwise) would be allocated to the survey officer who is responsible for the patch covering the location of that job. However, the survey officer may already have other urgent jobs in hand and so could not complete the new, urgent job in time. In which case, the central controller (normally the team manager) telephones other survey officers to check whether they can organise their workload to take the new jobs. The decision is mainly based on the distance between their current location and the location where the new job should be performed. Ideally, the current traffic situation on the route between a candidate survey officer and the new job should be considered as well. The priority of the job the survey officer is currently performing is also an important criterion.

All of this process is currently manual, relies on contacting the survey officers via their mobile telephones, and can be quite time consuming.

14.4.2 Group Decision-Making Process Modelling

The process definition for the above decision problem is shown in Fig 14.5. There are two roles involved — job giver and job receiver. The job giver is an initiator role which starts the decision process and represents the interests of a worker who wants to re-allocate some of their jobs to other survey officers. The job receiver is a respondent role which reacts to requests from the job giver.

Figure 14.5 (a) shows the decision process of the job giver role. The whole decision process is activated when a ‘GiveJob’ goal is created. The first action is to set a decision policy for the process, in this case, the maximum distance of a team member from the job location and a time-out value for a colleague to respond to a proposal. This information is requested from the user via a graphical user interface (GUI) on the mobile device. If a policy is given by a user, the ‘GetBids_initiation’ component is used to perform a ‘send call for bids (CFB)’ interaction. After sending the CFB, the initiator component waits until the time out has expired for the request or until all responses have been received. The received bids are evaluated and used as inputs to the next intra-action — ‘evaluate bids’.

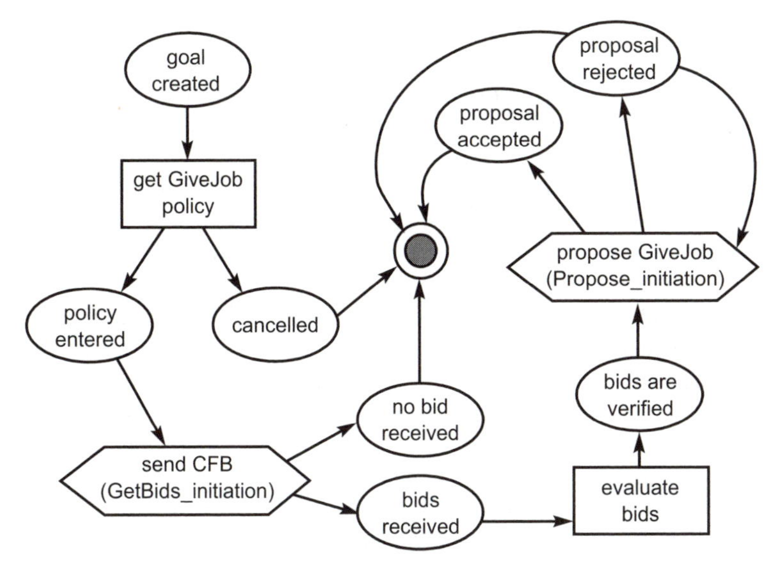

(a) Decision process definition for job giver (initiator) role.

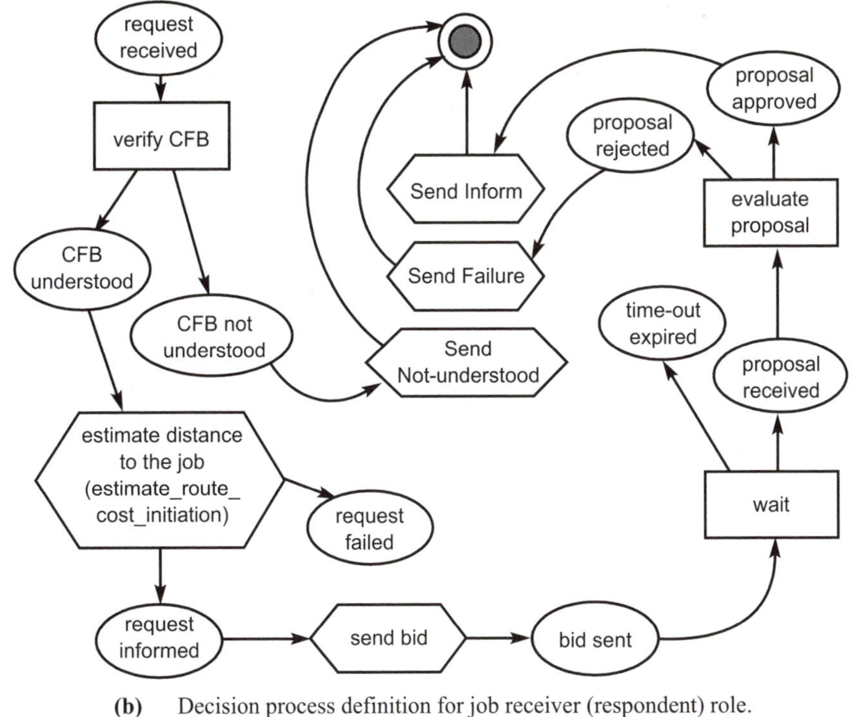

(b) Decision process definition for job receiver (respondent) role.

Fig 14.5 Goal plan for dynamic job re-allocation process.

This sorts the candidates according to their distance from the job (as contained in their bids), and the sorted candidate list is passed to the interaction 'propose GiveJob'. This interaction sends a propose message which contains an offer to delegate the job to each candidate in turn, until someone accepts the proposal or all team members refuse.

Figure 14.5(b) shows the decision process for the job receiver role. The process is activated when a request message arrives. The respondent component uses another two service components ('get current position' and 'estimate route cost') to calculate the distance between the job and the current location of the responding agent. Then a bid based on the distance is prepared and sent back to the initiator agent. If the agent receives a 'propose' message, the user associated with the respondent agent is presented with a GUI component showing proposal details for the GiveJob. If the proposal is accepted, the whole process is terminated. Otherwise, the proposal is passed on to the survey officer next closest to the job location.

In the decision process, most of the decision-making activities are performed by personal agents, which minimises the intervention of survey officers. The job giver simply specifies the job to be given and the decision policy to use. The job receiver is saved the bother of creating and sending a bid for job re-allocation. The personal agent autonomously creates this bid by using a GSCmas to get the distance from the current location of the survey officer to the job location. However, the critical decision whether to accept the proposal from the job giver is made by the survey officer receiving the proposal, ensuring that the computer system plays a decision support, rather than command and control, role within the team. This provides a degree of empowerment for each user — a key factor in maintaining effectiveness and increasing motivation [13].

14.4.3 Implementation

The mPower application was developed using Personal Java and was put through a trial by a team of survey officers in the UK. Each survey officer was equipped with a personal digital assistant (PDA) with network connectivity enabled by a PCMCIA GPRS (general packet radio service) card using an expansion sleeve for the PDA, or via Bluetooth to a GPRS-enabled mobile telephone. A GPS PCMCIA card was used to provide the location data. Secure access to the corporate intranet was via a GPRS virtual private network.

Figure 14.6 shows two screenshots of the mPower trial application. The left-hand image shows the screen presented to the job giver. Survey officers can list all their assigned jobs on the PDA and start a new decision process for re-allocation of work just by clicking the 'give to' button. When requested, the decision policy used by the personal agent (via the 'GiveJob options' dialogue box) must be specified. At the end of the process, the survey officer is notified whether the job has been accepted by another survey officer or rejected by every team member. The right-hand image in Fig 14.6 is seen by a job receiver. It shows who proposed the job and gives details, including the distance from the job receiver's current location to the job location.

Fig 14.6 Using mPower to support the dynamic job re-allocation process.

The field trial proved that agent technology works for distributed teams using state-of-the-art mobile devices, GPS and GPRS. The mPower application was used by a team of survey officers to support their day-to-day work. Example comments from the users in the trial provide a deeper insight into the effectiveness of mPower for mobile workforces. One survey officer talked of the usefulness of the GiveJob service:

He liked '... the ability of being able to send a job and for [mPower] to find the next nearest person, rather than have to ring around, especially when I am running two queues [patches] as I am at the moment.'

Another survey officer highlighted an interesting human/computer interaction (HCI) [14] issue:

'... it is all too easy to refuse the job at the press of a button. When you have someone you know on a phone line it is human nature to be more likely to say yes, but if it is a machine that will not say anything back if you press no, then I suspect that is what will happen in most cases.'

This confirms the need for further research into how to transfer the normal social pressures present in 'real' contact (e.g. a person wanting to improve their conversational partner's perception of themselves) to computer-mediated contact.

14.5 Discussion

The approach presented here provides some useful advantages when developing a GDSS for mobile teams.

Firstly, from a DSS perspective, multi-agent technology enhances the autonomy of a GDSS. Mobile workers spend the majority of their working day in the field, and as a result are frequently not able to access the computer-based information they need to do their jobs effectively. Autonomous agents can automatically execute some decision-making activities according to the workers' preferences, which minimises the interruption of a group decision process and facilitates rapid group decision making.

Secondly, from a software engineering perspective, the use of reusable GSCmas modules and group decision-making process modelling gives the GDSS a great deal of flexibility. The developer can define a new decision-making process, using existing GSCmas modules, which is then interpreted and executed by a user's personal agent.

14.6 Summary

In this chapter, mPower has been proposed to support asynchronous location-based group decision making in mobile teams. mPower is based on collaborating multi-agents which automate a group decision-making process by creating a group decision session, negotiating with other agents, and collecting information needed for the decision-making on behalf of their users. Particularly, the personal agent utilises a group of predefined generic service components. If at design time the personal agent needs various types of information, the relevant service components are installed. A group decision-making process can be modelled by specifying the reusable service components in each phase of the decision process. Lastly, the trial of mPower by a team of survey officers demonstrates that this technology is applicable in real world scenarios where a personal agent negotiates on behalf of the user.

A number of benefits arise from using this system:

- increased response time of job negotiations by not relying on user input when dealing with computer-readable contextual information such as current location;
- reducing the work required for a survey officer to complete their task, for example, shortening the time spent arranging jobs over the telephone;
- reduced communications costs.

Typically, a manual equivalent of the GiveJob service would involve three one-minute peak-rate calls over a GSM network — for a large corporation this would cost roughly £0.15 [15]. The data needed for the automated GiveJob service with three participants (including login and job download overhead) was roughly 6 kb, costing £0.007 — a twentieth of the 'manual' method. With a very conservative estimate that at least one GiveJob process happens once a week, we can estimate that, for a 25 000-strong workforce, an enterprise would save at least £60 000 per year by using the GiveJob service.

References

1 Klusch, M. (Ed): '*Intelligent Information Agents: agent-based information discovery and management on the Internet*', Springer, Berlin (1998).

2 Maes, P.: '*Agents that reduce work and information overload*', Communications of the ACM, **37**(7), pp 31-40, ACM Press (1994).

3 FIPA — http://www.fipa.org/

4 Role modelling — http://ootips.org/role-object.html

5 Bellifemine, F., Poggi, A. and Rimassa, G.: '*JADE — a FIPA-compliant agent framework*', Proceedings of PAAM'99, London, pp 97-108 (April 1999).

6 Collis, J., Ndumu, D., Nwana, H. and Lee, L.: '*The Zeus agent building tool-kit*', BT Technol J, **16**(3), pp 60-68 (July 1998).

7 Adorni, G., Bergenti, F., Poggi, A. and Rimassa, G.: '*Enabling FIPA agents on small devices*', Lecture Notes in Computer Science, **2182**, p 248, Springer-Verlag (2001).

8 Caire, G., Lhuillier, N. and Rimassa, G.: '*A communication protocol for agents on handheld devices*', Workshop on: '*Ubiquitous Agents on Embedded, Wearable, and Mobile Devices*', held in conjunction with the 2002 Conference on Autonomous Agents and Multiagent Systems, Bologna, Italy (2002).

9 Khoshafian, S. and Buckiewicz, M.: '*Introduction to Groupware, Workflow, and Workgroup Computing*', John Wiley & Sons Inc (1995).

10 Turban, E.: '*Decision Support and Expert Systems: Management Support Systems*', Macmillan, New York (1990).

11 Peterson, J. L.: '*Petri Net Theory and the Modelling of Systems*', Prentice-Hall Inc (1981).

12 Interaction Design Patterns — http://www.pliant.org/personal/Tom_Erickson/InteractionPatterns.html

13 Mehandjiev, N. and Odgers, B. O.: '*SAMBA — agent-supported visual interactive control for distributed team building and empowerment*', BT Technol J, **17**(4), pp 72-77 (October 1999).

14 Dix, A., Finlay, J., Abowd, G. and Beale, R.: '*Human-Computer Interaction*', Prentice-Hall Inc (1993).

15 O_2 Products — http://www.o2.co.uk/business/productsservices/

15

A MILLION SEGMENTS OF ONE — HOW PERSONAL SHOULD CUSTOMER RELATIONSHIP MANAGEMENT GET?

N J Millard

15.1 Introduction

The following story illustrates one of the dilemmas of personalisation and customer relationship management.

After booking an Australian holiday on-line from the UK, in Australian dollars, David tried to pay for his travel insurance. His credit card was refused. Annoyed, he rang his bank only to be asked if he knew that his card had been used to transact business in Australia. David said he knew this, he'd done it! He didn't understand why his card had been blocked. The card block was taken off and, although David felt comforted that his card was secure, he also felt annoyed that he hadn't been consulted before the card had been blocked.

Advances in data-mining technology and behaviour profiling mean that credit card companies have been able to cut down on fraud. However, on the occasions when behaviour is not fraudulent, this proactivity can have a detrimental (and in this case, embarrassing) effect on the customer relationship.

Customer relationship management (CRM) is built upon the foundation of a single definitive view of customers which spans functions, channels, products and customer types and drives every customer interaction. CRM purports to recreate the 'traditional corner shop' experience to millions of clients. At Capital One Bank, for example, they claim that they '... no longer have customer segments, we have a million segments of one'. CRM's lifeblood is the ability to deploy knowledge at the right time, in the right format, to the right person. Technology is at the heart of this, but technology alone is an enabler of CRM, not a guarantee of a good customer experience.

Fluss et al [1] define the role of technology in CRM as:

- capturing customer data across the enterprise;
- consolidating all internally and externally acquired customer-related data in a central database;
- analysing the consolidated data;
- distributing the results of that analysis to various customer touchpoints;
- using this information when dealing with customers via any touchpoint (e.g. call centre, face to face, Internet).

To customers the only part of CRM that actually matters is that their information is put to good use. CRM should enable companies to deliver a superior level of service. Unless this customer knowledge can be translated into a simple, personalised and easily digestible format, wherever and whenever it is needed, a differentiated service is unlikely to be offered.

15.2 How Personal Should CRM Get?

In an increasingly commoditised market, companies are looking to differentiate through their service surround. Delivering a tailored service is an attractive solution to this problem because the resulting relationship implies increased customer loyalty and, therefore, increased share of wallet. However, this comes from a company's desire to retain and acquire customers.

From a customer perspective, things are a little different. One of the fundamental assumptions of CRM is that customers actually want to develop a relationship with all the companies with which they do business. A McKinsey survey [2] of consumer attitudes to CRM revealed that customers do not want relationships, they just want decent service. Unless customers see a benefit in building the relationship, they tend to want to disclose as little as possible about themselves. Fisk [3] has claimed that '... today's blind rush to build relationships with customers is both unnecessary and unsustainable. It is not what customers want and is often bad for business. It is usually motivated by the desire for short-term revenues or internal efficiencies, but instead encourages commoditisation and promiscuity'.

What customers look for in a company in terms of customer service has been the topic of much research, but it boils down to a number of basic tenets [4-6]:

- remember who I am wherever and whenever I contact you;
- don't waste my time;
- treat me fairly and honestly;
- help me achieve my goals;
- be flexible to my needs;

- keep things simple;
- make it easy for me to do business with you;
- make information accessible to me;
- give me value for my money;
- make me feel secure in transacting business with you.

In a survey by the Direct Marketing Service [7], being treated as an individual was cited as being of primary importance by 36% of the customer sample. This figure rose to 40% for finance services and 46% for travel services.

Customers are starting to choose whether they allow companies to hold a full picture of their behaviour or remain a shadowy figure in the background, a presence without an identity. There is a growing suspicion among customers that revealing who they are, by giving out personal details, will create more problems than it solves. Orwell's concept of 'big brother' [8] and Bentham's panopticon [9], which both imply the use of personal data for centralised control (whether corporate or government), pervade the public consciousness. Linked with this is an increasing public awareness that consumer data is extremely valuable. The industry that has sprung up around selling company data lists was valued at £2 billion in the UK alone last year [10].

Organisations who are hungry for customer data are creating a general population who, tired from the barrage of junk mail/SMS/e-mail, are starting to become more cynical about corporate motives. Furthermore, Clive Humby [2] suggests that corporations have trained customers to be fickle.

Customers react badly when companies abuse customer information. Surveys consistently show that consumers believe that loyalty should be rewarded with lower prices [11]. However, practical implementations of the use of loyalty information to differentiate pricing have proven extremely unpopular. Amazon's attempt to tailor prices for different customers for the same DVDs nearly caused significant damage to their reputation.

To work effectively, personalisation should not be used to sell products but rather to better understand and serve the customer. Supported by proposed changes in legislation, the concept of 'permission marketing' [12] means that customers opt in to personalised service rather than the traditional approach of opting out.

If a customer does choose to have a relationship with a company, the fundamental building block is trust. Without trust, the future of any relationship is limited. A customer is unlikely to reveal any personal information to a company that it does not trust. Yet trust is more than security of a transaction. It is about partnership. If a customer chooses to share information with a company, then that company should respond with actions that create mutual benefit.

An integral part of trust is control. If customers feel that they are no longer in control of their own personal data they are less likely to disclose anything to anyone. In a poll by Harvard Business Review [13], 42% of Web users did not want to have sites personalised for them. Most valued filtering of the product selection based on their profile but they overwhelmingly wanted to have control of the filter. Interactive digital television has recently come under fire for their data collection and personalisation practices. Personal digital video recording systems, such as TiVo, build profiles of viewers' preferences and then automatically record programmes to profile. The viewer has no control over what is recorded but does have control over what they choose to watch. This has been the subject of a backlash from viewers who realise that they are constantly under surveillance [14].

Further public worry has been provoked by the partnership between Microsoft and NDS (the Rupert Murdoch owned satellite TV technology company) who are developing technology to enable detailed profiling across subscribers' Internet and TV watching behaviour [14].

15.3 The Practice of Personalisation and What It Means to Customers

One of the key things in personalisation is to find out what personal information the customer wants a company to remember. Then the company must seek the customer's permission to use this information. Personalisation is not just accumulating data on individuals but acting upon it. The three steps to personalisation are:

- interact — talk to customers, establish a dialogue, understand their needs;
- invent — use the data differently;
- personalise — do things uniquely.

Invention is the key to differentiation of service. Best practice companies do a number of things with personalised CRM that other companies simply do not. They do not simply act as a black hole sucking in data and producing nothing back. As Clive Humby says '... to truly know your customer there is one magic ingredient you need to add to your CRM data. It's imagination ...' [2]. Individually tailoring products and services to suit customers requires some smart thinking. Three elements need to be personalised when interacting with customers [15]:

- content (what the customer receives);
- context (how the customer receives it);
- contact (the way in which they are handled).

The concept of a customer 'loyalty' card, as exploited by numerous retail outlets, is to gain customer data by providing benefits to customers as part of a 'club'. Thus, purchase data is not the driver of the process, rather it is what is done with that data and how it is harnessed to influence the customer experience that is the key. This is a data-rich process, but there is little evidence that it achieves greater loyalty or share of wallet. The decision to invest in personalisation usually centres around attracting and retaining new customers.

Extending the idea of customer retention, Payne et al [16] introduced the idea of a 'loyalty ladder' where customers develop from low-spending prospects, to customers, clients, supporters and advocates as the relationship develops. The implication is that an emphasis on developing and enhancing customer relationships moves customers up the ladder over time.

However, more recent research has pointed towards a more subtle relationship between loyalty and profitability. Customers do not develop a propensity to spend more with a company the longer that they have been with them. Loyal customers are not necessarily cheaper to serve, less price sensitive or particularly good advocates for the company [11]. Research has shown that newer customers often act as better advocates for the company than more loyal ones [17].

This calls into question the economics of loyalty card schemes. A loyalty scheme that gives a 1% discount on gross margins of 20% requires a 5% increase in sales in order to make a profit [18]. This has caused a number of high profile companies, notably Safeway, to withdraw from them altogether. When developing personalisation strategies, companies must consider how best to manage the value of the customer over time.

Companies who have harnessed a personalised CRM strategy, such as Amazon, First Direct, Capital One and Tesco, have all largely focused on the most cost-effective channel that they have to deliver this. Their approaches are very different but they are all using their customer data in different ways to tailor their customer interaction. Amazon has pioneered a personal approach to on-line interaction using the buying patterns of their customers. Both Capital One and Tesco use variations on a data-driven approach. First Direct simply use people to personalise.

15.3.1 Personalisation On-Line — Amazon Case Study

In a survey of Web users, Peppers and Rogers [19] found that:

- customers want to be treated as individuals;
- they not only like personalisation on the Web sites they visit, they expect it and get annoyed when they do not get it;
- they understand that personalised Web sites require them to provide personal information, and they are willing to provide that information as long as they trust the company to protect their privacy.

The survey showed that 73% of Web users wanted a Web site to remember basic information about them (after permission has been given to do so) and 50% are willing to share personal data to get a personalised on-line experience. 62% are annoyed when a Web site asks them for personal information they have already given. Yet 59% of users said that they would not share information unless there was a recognised statement of privacy. Amazon are one of the finest practitioners of on-line personalisation.

Amazon's Web site uses a personalisation technology called 'collaborative filtering'. Once you have made a purchase with Amazon.com, it recommends a new book by comparing your purchase history with those of fellow book buyers who have bought that book [20]. This is a method echoed by a number of on-line retailers and also has uses to power recommendation engines for entertainment and television viewing.

Amazon's personalisation engine offers more than simply recommendation. Once it has recognised the customer using the cookie stored on the computer, it brings in many of the basic tenets of customer service. Not least it saves the customer's time and effort with the 'buy it now' button. This means that the customer, once registered, simply has to make one mouse click to buy rather than repeatedly having to input personal or credit card details.

This, as well as the ability to deliver products quickly, has created enormous on-line brand loyalty for Amazon (who also had first move advantage) over more established 'clicks and mortar' brands such as Barnes&Noble.com and Waterstones.com. Controversially, Amazon even patented their one-click shopping system [21] to deprive their major competitors from delivering their innovative spin on customer relationship management.

Yet the most remarkable aspect of the Amazon strategy is that they have created a highly personalised CRM experience without the need to speak to a human being. A simple approach to user interaction on their Web pages creates the illusion of a one-to-one relationship. The assertion that 'all that people want is good service' is fulfilled here on a low-cost channel. Their personalisation strategy has developed people as 'Amazon shoppers' to the exclusion of other on-line book vendors — yet Amazon book buyers will probably still shop in a book store on the high street. Amazon does not replace the role of the high street retailer, it complements it.

However, the recommendation engine model does not necessarily universally translate. Tesco.com discovered that their most frequently purchased on-line product was, surprisingly, bananas. A recommendation engine that they developed looking at anticipating purchasing habits often recommended that customers should buy bananas on the basis of what was in their shopping basket [22]. It took no account of the fact that the customer did not buy bananas because he or she did not like them!

15.3.2 Using Information to Personalise — Capital One Bank and Tesco

Capital One Bank has become master of the apparently paradoxical art of mass customisation. The company uses a mathematical strategy called 'optimisation' to form microsegments that allow it to essentially create millions of segments of one — personalising both product and experience for each of its customers. Optimisation is a process of objectively and mathematically allocating the customer management resources available to them across multiple channels, business constraints, and scenarios in order to determine the right offer at the right price for the right customer, through the right channel at the right time.

Its optimisation software can analyse a customer's likely preferences based upon past orders and spending in real time and target marketing campaigns accordingly. Capital One's data warehouse holds basic information on 120 million households, detailed data on 10 million customers and performance/profitability information on more than 4000 product, pricing and feature combinations. The company uses the information to predict consumer behaviour and then matches prospects to more than 300 credit cards with various terms and fees.

Bearing the brunt of this complexity is its call centre that has to cope with 217 call types. As a result it has 54 advisor groups and 14 switches plus an intelligent call-routing mechanism which is driven by the underlying optimisation model. The result is a proactive experience where customer service advisors are able to use their discretion to cross and upsell products based upon the customer profile and call type. For example, if the customer reports a lost or stolen card, the system might suggest that the advisor sell the customer a credit card registration service. Advisors also adapt the system by reporting customers' reactions to offers so that the company can continuously refine its portfolio.

Capital One has achieved higher success in marketing campaigns with this initiative and has seen a 10% growth in their customer base. In terms of the basic tenets of customer service, Capital One remembers who people are and is flexible to their needs, while hiding complexity embedded within the technology. Yet the principal reason for their success is that they used personalisation to drive their contact strategy both with the customer and with their employees. This includes the way that call centre advisors are measured. Marge Connolly, Capital One operations director, suggests that: 'Since our philosophy is to use every contact to build a customer franchise, we want to handle calls quickly but not as a way to drive down costs... The criteria have changed from a strict adherence to a set of rules for what should happen during a phone interaction to one that is more subjective.'

Capital One has also increased its use of customer surveys and has specially trained some of its Web site employees to be able to interact more effectively with Web users, all in an attempt to make sure that employee training matches customer needs.

Similar to Capital One, Tesco has built their whole business strategy on customer data. They practise the art of 'reverse marketing' [2] using customer pull rather than marketing push. Clive Humby, who developed the loyalty card scheme, explains that '... the data allows Tesco to create marketing plans for groups of customers, finding products for them rather than customers for products'. Tesco believes that loyalty is earned by being loyal to its customers. Rather than being simply a discount mechanism, like most loyalty cards, it uses pattern-matching techniques to effectively mine the millions of pieces of data that come in via its stores' tills. Tesco makes use of seven fundamental pieces of data — customers' lifestage, shopping habits, what they buy, response to promotions (including competitor promotions), primary channel, brand advocacy, and cost to serve/profitability — to steer pricing, store development and product development (including brand extension products such as personal finance). Tesco also uses clubs such as the Kids' Club to further identify its customers as 'Tesco customers'. Its promise is based upon improving service and it pledges never to pass this valuable customer information to anyone else.

This data-driven approach lends itself to the Internet as well as to physical stores. As a result, Tesco has become the world's most successful on-line grocer. It has blurred the line between the on-line and physical worlds by effective use of on-line shopping lists. It found that customers tended to abandon 'shopping carts' because of the problems navigating around an A to Z of products. In response, it offered first-time visitors to the site a suggested shopping list based upon their clubcard data drawn from shopping patterns in stores.

Tesco has put an innovative spin to the way that it delivers experience, through responding to customer need by gaining a solid understanding of what customers want from the company through loyalty card data.

15.3.3 Using People to Personalise — CRM and First Direct

CRM is not purely a technological experience, as shown by First Direct who has taken a subtly different, low-tech approach to personalisation. It has taken the revolutionary step of using people to personalise and train its telephone-based agents to be extremely sensitive to customer need. Their training techniques emphasise customer empathy and mood matching which serve to create a highly personalised experience through rapport-matching on the telephone. There is little personalised technology supporting this, just sufficient information from a customer database and weeks of training on empathy and customer care techniques that ensure that customer service advisors can truly match the customer's mood and need.

One of many classic First Direct customer experience anecdotes depicts an advisor talking to a customer about a problem with their finances, noticing it was his birthday from his security validation field and, at an appropriate moment, bursting into 'Happy Birthday to You'. This simple yet personal touch adds up to a powerful customer experience which plays to the natural strengths of humans (rapport, relationship building, complex problem solving), and results in one of the UK's highest customer satisfaction levels for banks.

15.4 Future Directions for Personalised CRM

Looking to the future, there are a number of trends that are the focus for strategic and technological development, of which four are discussed here.

15.4.1 Understanding Customer Value and the Economics of Strategic CRM Investment

The economics of personalised CRM strategies are increasingly under scrutiny as companies struggle with understanding the value of their customers and the return on investment that they get from different approaches to serving their customers [23]. Supplying a highly personalised and expensive channel to a customer who spends very little on your services is not beneficial to any company's bottom line. In terms of the economics of delivery, the Internet is likely to become the focus of these developments, since it is the most easily personalised at the lowest cost.

15.4.2 Building More Sophisticated Business Intelligence Tools to Analyse and Utilise Customer Data

Business intelligence and the development of ever more sophisticated techniques of data mining provide significant ways forward in terms of allowing companies to use customer data to deliver differentiated experience. With the Data Protection Act preventing detailed manipulation of an individual's data, more refined pattern recognition, using statistical modelling and fuzzy logic without the need to store individual information, is one promising research focus. These techniques can calculate the probability that some future event will happen based upon statistical patterns in observed consumer browsing behaviour. This can translate into providing a personalised user experience which could potentially allow Internet pages to dynamically build themselves according to whether the customer has transacted business on the site before and also to their potential to purchase.

15.4.3 Reducing Information Saturation for both Customers and Customer Service People

With an increase in the level of data analysis and sophistication in delivering personalisation to the customer, there is a danger that the level of complexity at the customer interface can start to adversely affect the service delivered. Attention must be given to the design and usability of information provided to both the customer and the front-line employee. As Amazon has proved with their Web site design, simple is best.

Knowledge management tools within the call centre must be able to supply customer information in a timely and meaningful way so that customer service advisors can also use customer information to serve. By providing a personalised way for the service advisors to access information, transactions with customers can be both more accurate and more effective.

The tempting solution is to provide the advisor with seven different windows of customer information in seven-point font on their screens. Having a wealth of customer information on screen while trying to remember complex rules and procedures, as well as having a coherent conversation, stretches the capacity of human working memory and processing ability [24]. By designing screens to cope with the limitations of the human brain through effective and intuitive user interface design [25], advisors can be equipped with tools to deliver a personalised customer experience more effectively and efficiently.

15.4.4 Understanding Customers' Needs for Personalisation

It is a fine line that CRM strategies must tread in terms of privacy and customer trust. The Egg metaphor of 'dancing with customers' illustrates a view that companies must understand and abide within the acceptable 'personal distance' rules that a customer sets. If they overstep the boundary, their reputation is on the line. Tesco's strategy of creating customer pull rather than marketing push allows companies to learn, and adapt to, customer need based upon actual behaviour. Capital One's approach allows them to experiment with customer data and model customer responses in a benign environment before they risk a strategy in the market.

Using data to better understand customer need, rather than simply trying to sell them something, creates a far greater mutual benefit and, ultimately, a better customer experience.

15.5 Summary

Ultimately, the future of personalised CRM is dependent on non-intrusive, mutually beneficial and cost-effective strategies delivered through appropriate customer touchpoints. This requires recognition that CRM is more than just deployment of technology — it is a fusion of strategy, process and technology. Companies need to consider what customers want from them and whether the personalised solution that is being proposed reaps financial benefits. Building a million segments of one may work for some companies but others need to consider what kind of relationship their customers want from them and what they (and their customers) can gain from the data that they gather.

References

1 Fluss, D., Amuso, C., Hope-Ross, D. and Ross, C.: '*Internet-based customer service: miracle or migraine*?', Gartner (September 1999).

2 '*Clive Humby Interview*', Customer Management, **10**(5) (September/October 2002).

3 Fisk, P.: '*The rise of the customer gateway*', Customer Service Management (November/December 1999).

4 Seybold, P.: '*Customers.com*', Random House (1998)

5 Gober, M.: '*The 12 things customers really want*', Customer Management, **8**(1) (January 2000).

6 Ash, I.: '*The customer experience*', Internal BT report (1993).

7 Personalisation Conference: '*How to succeed through profiling*', Privacy and Permission Marketing, Business Intelligence, London (November 2001).

8 Orwell G: '*1984*', Penguin (1949).

9 Bentham, J.: '*The principles of morals and legislation*', Prometheus (1780).

10 Lawrence, F.: '*Checkout at the data supermarket*', The Guardian, Big Brother Supplement (September 2002).

11 Reinartz, W. and Kumar, V.: '*The mismanagement of customer loyalty*', Harvard Business Review (July 2002).

12 Godin, S.: '*Permission Marketing*', Simon & Schuster (1999).

13 Nunes, P. F. and Kambul, A.: '*Personalisation? No Thanks*', Harvard Business Review (2001).

14 Wells, M.: '*Watchers and the watched*', The Guardian, Big Brother Supplement (September 2002).

15 Vandermerwe, S.: '*Taking customers personally: the imperative for driving corporate growth*', Personalisation Conference, Business Intelligence (November 2000).

16 Payne, M., Christopher, M., Clarke, M. and Peck, H.: '*Relationship Marketing for Competitive Advantage*', Butterworth-Heinemann (1997).

17 Ranaweera, C.: '*Investigating the drivers of customer retention*', PhD Project, Judge Institute of Management Studies, Cambridge University (2001).

18 '*Is CRM really effective*', Ibhar Business Partner Program, White Paper (2000).

19 Peppers, D. and Rogers, M.: '*Wanting it all: give us personalisation and privacy*', Inside 1 to 1 (April 2000).

20 Weinstein, A. and Johnson, W. C.: '*Designing and Delivering Superior Customer Value*', CRC Press Ltd (1999).

21 '*Amazon.com: A Business History*' (April 2002) — http://faculty.washington.edu/sandeep/textbook/

22 Arnold, P.: '*Using operational excellence as a competitive advantage*', Competing in the Age of Customer Control, Benchmark (December 2001).

23 Blodgett, M.: '*Masters of the customer connection*', CIO Magazine (August 2000).

24 Sweller, J.: '*Cognitive load during problem solving: effects on learning*', Cognitive Science, **12**, pp 257-285 (1988).

25 Millard, N., Hole, L. and Crowle, S.: '*From command to control: interface design for customer handling systems*', Procedures of INTERACT 98, Chapman & Hall (1998).

INDEX

Accuracy 16-19, 34-39, 51-53, 188-192
Adams P M 43
Advertising
  banner 162-170
  broker 162-171
  on-line 162-169
  personalised 77, 161-171
Affinities 3
Altitude 17-20, 52
Angle of arrival 187
Anonymity 8, 40, 170
Appleby S 129
Applications 15, 20-21, 24-28, 76-77, 167-168, 192-194
  consumer 24
  hotel booking 192-193
  interface 72
Application server (AS) 83, 120
Architecture for Location-Based Service, *see* Location-based service
Ashwell G W B 43
Authentication 8, 65, 88

Baxter R 43
Bilchev G 15, 161
Billing mechanisms for LBS, *see* Location-based service
Bluetooth 118, 206
Bookmarks 72-75, 103-104
Buckland M A 197

Call processing language (CPL) 85
Capability and preference information (CPI) 119
Case S 173
CellID 18-19, 25-28, 36-39, 51
Cellular positioning 18
Collaboration 95-97, 177-178
Collaborative filtering 3, 216
Collingridge R 59
Common presence and instant messaging (CPIM) 89
Communities 76-78, 173-183
  of practice 95-97
Community-based applications 26
Composite capability/preference profile (CC/PP) 84-85, 119-125
Consumer applications, *see* Applications
Content adaptation 85, 116, 123-124
Context 3-6, 55-57, 105-111, 133-135, 214
Cookie 75-76, 98-100
Co-ordinate systems 20
Co-ordination Group on Access to Location Information by Emergency Services (CGALIES) 35
Crossley M 93
Customer relationship management (CRM) 3, 132, 211-222
Customisation 7-8, 88-89
Customised application mobile enhanced logic (CAMEL) 82-83

Data-mining 98
Data Protection Act 68, 219
Decision-support system (DSS) 199
 see also Group decision-support system
Device personalisation 115-126
De Zen G 81
Differential GPS (DGPS) 17-18
D'Roza T 15
Dynamic job allocation 204

E112, *see* Emergency
E911, *see* Emergency
Edit decision list (EDL) 155
Emergency
 E112 28, 31-34
 E911 28, 31-33
Enhanced-observed time difference (EOTD) 36-39
EU 31-35, 40, 51-52
eWallet 65
Example-based machine translation 141
Extensible
 markup language (XML) 21-23, 56-60, 85-86, 110
 stylesheet language (XSL) 124
 style-sheet language transformations (XSLT) 22, 84
Extranet 59, 178

Federal Communications Commission (FCC) 33-34, 186
Fleet management 24-25
Font 133-135
 see also TrueType
Foundation for Intelligent Physical Agents (FIPA) 198-201

Gabrielle 46
Galileo 51-52
Gardner M R 115
General packet radio service (GPRS) 118, 206-207
Generic user profile (GUP) 86-88
Geographic information system (GIS) 21, 40
Geography mark-up language (GML) 21-22
Gillies S 115
Global Positioning System (GPS) 15-18, 51-52, 186-193
Global system for mobile communications (GSM) 18, 39, 43-50, 82
 cell location 60, 64
Glyph 133-135
Graphic user interface (GUI) 89, 131, 204-206
Group decision-making process
 definition 201
 modelling 204-206
Group decision-support system (GDSS) 197-199, 207-208
GUIDE, Government User Identity for Europe 66-67

Hare S 173
Hoh S 115
Home
 environment (HE) 82
 subscriber server (HSS) 54, 83
Hosting the profiles 121
Hotel booking, *see* Applications
Human/computer interaction 60, 207
Hypertext transfer protocol over secure socket layer (SSL) (HTTPS) *see* Protocol

Identity 8, 66-76, 165-166
 management 61-65, 78
Input method editor 136
Intelligent
 agent 9, 197
 see also Software agent

network 55, 217
Interest 5-9, 57, 67, 73-75, 94-110, 174-176
Inter-lingua 140
Internationalisation 130-132
Intranet 59, 178, 206
IP multimedia subsystem (IMS) 83

Java 2
micro edition 198
standard edition 198

Kerberos 65
Kings N J 93
Knowledge management 5, 93, 220

Latitude 17, 20, 186-192
Lee H L 197
Liberty Alliance 7, 65-66, 78, 171
Lightweight directory access protocol (LDAP), *see* Protocol
Localisation 130-133
Location
in Emergencies 31-32
Interoperability Forum (LIF) 40, 49, 55, 57
requirements for terminals making emergency calls 44
technology for emergency services 36-39
Location-based service (LBS) 15-27, 43, 185, 198
architecture 53-55
billing mechanisms for 27
standards 43-57
Logistics 193
Longitude 17, 20, 186-192
Loyalty card 3, 168-169, 215-218

Machine translation 136-143
Marston D 161
Media 69, 115-117, 123, 147-152, 158-159
asset management 151
mark-up 152-153, 156-157
Metadata 6, 120, 151-156
Microsoft
.Net My Services 65
Passport 65, 70
Millar W 185
Millard N J 211
Mobile location
computer (MLC) 185-186, 188-194
protocol, *see* Protocol
MPEG7 152
Multi-agent system (MAS) 197
Multipath analysis 187

Navigation mark-up language 22
.Net My Services, *see* Microsoft
Newbould R 59

Object-based media 147, 159
Observed Time Difference of Arrival (OTDOA) 51
Ohtani T 173
On-line advertising, *see* Advertising
Open
agent middleware (OAM) 173-179
Mobile Alliance (OMA) 46, 48-49
service architecture (OSA) 82-83
OpenType 135
Oracle Spatial 23

PCMCIA 206
Permission marketing 213
Personal
agent framework (PAF) 173-180
digital assistant (PDA) 9-10, 115, 206
Java 198, 206
see also Java 2
profile 4-6, 10-12, 65, 71

service environment (PSE) 82, 89
Personalisation 1-8, 58, 67-72, 81-85, 89-90, 94-96, 116-117, 155-156, 173-174, 214-220
Personalised
advertising, *see* Advertising
alerting 76
Platform for privacy preferences (P3P) 6, 64, 73, 170
Point-of-interest exchange (POIX) 21-23
Portal 4-5, 57, 84, 97, 163-164
Preferences 4-9, 65, 75, 81-89, 119-120, *see also* User
Presence 56-57, 88-89
Privacy 5-6, 40, 53, 63-64, 67-70, 161-162, 169-171, 175-176
policy 6, 68, 162, 170-171
Profile hosting 59-60, 67-71, 99
Profiling 59-77, 85-88, 97-104
Protocol
hypertext transfer over secure socket layer (SSL) (HTTPS) 70-72
lightweight directory access 70-72
mobile location 40, 54-57
session initiation 83-85
simple object access 57, 60, 70-72
wireless application 48-50, 56-57, 87-88, 115-122
Pseudonymity 170-171
Public
key cryptography 64
safety answering point (PSAP) 32-40

Rayleigh 108
Reality TV 149, 156
Redknee 57, 188
Resource description framework (RDF) 85-86, 119-122
Rights management 9-11, 76
Roadside vehicle recovery 193-194
Roles 73-74, 102, 178, 199
Russ M 147

Salmon P H 31
Scott J R 93
Searby S 1
Search engine 95, 109-111, 132
Security policy 88
Selective availability 16
Serving call session control function (S-CSCF) 83
Session initiation protocol (SIP), *see* Protocol
Shepherdson J W 197
Simple object access protocol (SOAP), *see* Protocol
Single sign-on 4, 63, 65-66, 171
SKiCAL 22-23
Software agent 105-173
see also Intelligent

Targeted marketing 2-6
Taxonomy 104
Text
rasteriser 133-135
summarising 105
Thint M 173
Third Generation Partnership Project (3GPP) 44-53, 82, 89
Touchpoint 212, 220
Transcoding 123-124
Translation 77, 129-132
memory 132
see also Machine translation
TrueType 135
see also Font

Unicode 134-135
Uniform resource identifier (URI) 120-122
Universal Mobile Telecommunications

System (UMTS) 39, 44-54, 82
Urban canyon 18
User
agent profile (UAProf) 84-85, 119-122
location 50-51, 88
preferences 81, 84, 89, 119
see also Preferences
profile 6-7, 59-60, 65, 75-76, 81-83, 86-88, 96-98, 154-155, 161-166, 171-176
see also Profiling

Value-added service provider (VASP) 82-83
Verinymity 170
Video editing 149, 157
Virtual home environment (VHE) 82-83, *see also* Home

Web Services 5, 12, 60-62, 65, 70, 76-78, 195
description language (WSDL) 60, 70, 110, 195
'Where's my nearest...?' 25
Williams D 147
Wireless Application Protocol (WAP), *see* Protocol
Workforce safety 25
World Wide Web Consortium (W3C) 49, 73, 84, 119-120

XML, *see* Extensible
Zipf's law 2-3